# Epig

## IN THE

# Age of Twitter

## Pop Culture and Modern Science

# Epigenetics

## IN THE

# Age of Twitter

## Pop Culture and Modern Science

*Gerald Weissmann*

BELLEVUE LITERARY PRESS

*New York*

First Published in the United States in 2012 by
Bellevue Literary Press, New York

FOR INFORMATION ADDRESS:
Bellevue Literary Press
NYU School of Medicine
350 First Avenue
OBV 640
New York, NY 10016

Bellevue Literary Press would like to thank all its generous
donors, individuals and foundations, for their support.

Cataloging-in-Publication Data is available from the Library of Congress.

*Book design and composition by Mulberry Tree Press, Inc.*
Manufactured in the United States of America.
FIRST EDITION
1 3 5 7 9 8 6 4 2
ISBN 978-1-934137-39-0

Ann: *rien sans elle, toujours*

# Contents

She could not be sure, for the Machine did not transmit nuances of expression. It only gave a general idea of people—an idea that was good enough for all practical purposes. . . . Something "good enough" had long since been accepted by our race. Then she generated the light, and the sight of her room, flooded with radiance and studded with electric buttons, revived her. There were buttons and switches everywhere—buttons to call for food, for music, for clothing. There was the hot-bath button [and] the cold-bath button. There was the button that produced literature. And there were of course the buttons by which she communicated with her friends. The room, though it contained nothing, was in touch with all that she cared for in the world.[1]

—E. M. Forster, "The Machine Stops," 1909

It's as if Zuckerberg read E. M. Forster's famous rallying cry in *Howards End*, "Only connect," and took it literally. . . . (There's no chance that this actually happened. I asked Zuckerberg if he'd read Forster and got the spider stare. He'd never heard of him.)[2]

—R. Stengel, "*Time*'s Man of the Year," 2010

# Prefatory Note

THE ESSAYS IN THIS BOOK deal with modern science and the popular arts. There are, of course, real differences between art and science: in science if another lab replicates your work, you call to congratulate them; in the arts your lawyer calls. Another difference is that no real work of science has ever made anyone laugh out loud.

Arts and Sciences at one point did mean "Paintings and Stuff and Petri dishes and Stuff," to quote Zadie Smith.[1] But in the age of Twitter, Facebook and Buzznet, arts and sciences follow YouTube's charge to "Broadcast Yourself!" These days one click on Twitter separates Shawn Corey Carter (Jay-Z), a rapper who is CEO of Roc-A-Fella Records and who owns the NJ Nets, from Eric Kandel, our leading neuroscientist, who owns a Nobel Prize.[2] Walter Benjamin, the "popular wise guy,"[1] had it right: every reader has become a writer—or tweeter—in the age of electronic reproduction.

These essays play off one or another discovery of today's biological science (with epigenetics leading the pack) against one or another aspect of popular culture: film, television, sports, etc. (with Twitter et al. leading the pack.) We'll see what the molecular biology of aging has to do with Sarah Palin's hair, why a Nobel Prize for reproductive biology clashed with J-Lo's notion of fertility and why the name on Oklahoma State's football stadium (T. Boone Pickens) scuppered basic research on anthrax.

Twitter first. To learn what Twitter could do for me, I signed up a couple of years ago under the nom de silico "Franz Anton Mesmer" (1734–1815) or franzmesmer@twitter.com. Sure enough, I soon had half a dozen "followers." I told this to a psychiatrist friend of mine, who was astonished that anyone would follow the words of a dead hypnotist. I pointed out that this was the basis of organized religion, not to speak of Freudian analysis.

While Twitter is the latest effort of popular culture to see how little one can say to the largest audience, epigenetics is the latest twist on the historic nature/nurture debate in biology. Epigenetic changes are short-term heritable alterations in gene expression that are not due to mutations in DNA. An analogy: if DNA is a language composed of the letters A,T,G,C, epigenetic changes are like variations in a letter's font, i.e., C to **C**, or its context, i.e., C to ©. We've pieced together some of the machinery that accomplishes this trick, but we're still at sea as to how it all works.[3]

Darwinian evolution—the origin of species—takes place over eons: epigenetics prints short-term bar codes on our cells so that they remember where they've been and where they should be going. If they get lost, we call it cancer, obesity or diabetes. Since epigenetic changes can be induced by diet, drugs and the environment, there's a whiff of Lamarck about the whole notion.[3] Not to worry, though: until the genome crumbles, our skin will remain skin, our bones remain bones and apes won't turn into humans tomorrow. One trend is sure, though: nurture is gaining on nature in the old debate.

Epigenetic explanations have been extended to reach Woody Allen's second favorite organ, the brain. If the gonads (Allen's favorite) are the organs of classical evolution, the brain is the organ of personal and cultural evolution. Neurobiologists now believe that our brains evolve through a series of choices made by individual neurons over time. These choices are made under epigenetic control and, as in classical evolution, are forged by selective pressure with adaptive advantage.[4] Growth, learning and remembering take place in the *real* social network: imagine the very different selective pressures acting on the neural networks of young women growing up in Kabul or Minneapolis.

My essays touch on the social network we call Western civilization: Antoine and Marie-Anne Lavoisier, Emerson and Margaret Fuller, Gustav and Alma Mahler; Bernard Shaw and H. G. Wells. Experimental science, I'm convinced, is a crowning achievement of that civilization, and the fact that women now function as equal partners in science is a welcome sign of its continuing vitality. Several of the chapters in this book deal with the many women of science—from the martyred Hypatia of Alexandria, the first woman scientist, to the Nobel laureates Marie Curie, Christiane Nüsslein-Volhard, Elizabeth Blackburn and Carol Greider. The list speaks for itself.

Which brings me back to that business of laughing out loud. The first urbane female investigator of record was Dashiell Hammett's character,

Nora Charles. Her crack in the 1939 movie *Another Thin Man* rang a bell for women's liberation in the modern era:

"A Bacardi," says Nick, her husband and co-investigator, to the waiter in a Latin nightclub. He looks to Nora, who nods. Nick motions to the waiter and adds, "Two Bacardis." Says Nora with a straight face to the waiter, who'd ignored her, "I'll have the same." The waiter returns with four Bacardis. [5]

Cheers!

# Walter Benjamin and Biz Stone: The Scientific Paper in the Age of Twitter

| | |
|---|---|
| Maureen Dowd: | *If you were out with a girl and she started twittering about it in the middle, would that be a deal-breaker or a turn-on?* |
| Biz Stone: (dryly) | *In the middle of what?* |
| Maureen Dowd: | *Why did you think the answer to e-mail was a new kind of e-mail?* |
| Biz Stone: | *With Twitter, it's as easy to unfollow as it is to follow.* |

—Maureen Dowd, "To Tweet or Not to Tweet," 2009[1]

*For centuries a small number of writers were confronted by many thousands of readers. This changed toward the end of the last century. It began with the daily press opening to its readers space for "letters to the editor." And today . . . at any moment the reader is ready to turn into a writer.*

—Walter Benjamin, "The Work of Art in the Age of Mechanical Reproduction," 1931[2]

*All registered users are able to add Notes, Comments, and Ratings to any article. . . . Highlight the text to be annotated, and then click the "Add a note to the text" link in the right-hand navigation menu of the article. . . . Notes can be started at any point within the text, but for ease of reading we ask that you do not begin Notes in the middle of words.*

—Public Library of Science, 2009[3]

## The Sandbox of Ideas

AFTER WIKIPEDIA AND WIKILEAKS, are we now into Wikiscience? It's reassuring to read that our colleagues at the Public Library of Science have remained true to the integrity of the word, if not the sentence or thought. *PLoS One* has raised its banner for verbal integrity in a cheery commercial titled "*PLoS* Journals Sandbox: A Place to Learn and Play."[3] The new format, which permits instant interruption of online, formal scientific papers, is certainly in keeping with the temper of our time. Were this to have been the practice in old-fashioned print libraries, many of our journals would by now resemble Kitty Litter. And now there's Twitter.

In the Age of Twitter we've become accustomed to bell-tones and roving thumbs in every venue of human life. We call it social networking when we summon up Facebook, YouTube or MySpace—and it's no longer limited to teenagers. Twitter and the other social networks have been used by nearly one in five of online adults ages 25 to 34.[4] Nowadays, in the plenary sessions of national scientific meetings, one sees heads bowed in homage to the Holy Book of Face or tweeting to Twitter in fewer than 140 characters of text. Biz Stone, the founder of Twitter, explains:

> *Twitter is a service for friends, family, and co-workers to communicate and stay connected through the exchange of quick, frequent answers to one simple question: What are you doing?* [5]

Folks can't wait to tell you what they're doing. They've tweeted at funerals and inaugurations, texted en route to train crash and trolley disaster, so why not in lecture halls and laboratories, restaurants or concert halls?[6]

And as for science: what are we doing? Today, on screens large and small, every online scientific paper is just a cursor stroke away. That makes it possible, as Benjamin predicted, for any reader to turn into a writer. No surprise, then, that *PLoS* and other new-venture journals encourage us to adorn the digital text with notes and comments, blogs and tweets. Biz Stone brags that Twitter "will always be about providing access to a communication network through the lowest common denominator."[5] Right on to the *Public Library of Science!* How fitting it is that *PLoS*, the youngest kid on the block of reputable science journals, is out to compete in the sandbox of ideas.[3]

## Endangered Species of Print

It's no secret that scientific journals have been losing readers of their printed versions to the greater audience on the web. For many scientific journals, the number of "hits" they receive daily online is a factor or two greater than their monthly print circulation. That's true of the printed page in general: daily newspapers, news magazines and hardcover books are losing readers. Bookstores and publishers are going belly-up, and although books aren't being burned these days, they're being Kindled. In the words of John Kerry, newspapers are becoming "an endangered species."[7] The printed word still retains a good chunk of older devotees, but even these are as likely as their younger colleagues to prefer electronic to printed copies of their favorite journals.[8] The main threat to printed scientific journals is the cost of publication, since libraries alone can no longer support subscription costs.[9] The solution: having the authors pay, and pay enough to have their manuscripts published to support the editorial effort—and then some! Adapting Walter Benjamin's phrase, nowadays every author is ready to turn into a publisher.

This sea change in the way that information is handled and supported has worried many and frightened a few.[9] We might recall that scientific journals as we know them are relatively recent arrivals on the scene and have moved along paths trod by the general culture. Before there were any periodicals, Copernicus and Galileo, Vesalius, and Harvey did pretty well with monographs subsidized by court and/or learned societies. Science and publishing became professionalized at the dawn of the Enlightenment. The two oldest scientific journals on record are *The Philosophical Transactions of the Royal Society* (London) and the *Journal des Sçavans* (Paris), both founded in 1665. Originally filled with material of general interest for fellow citizens of the republic of letters, they soon morphed into publications that reported the most rigorous science of the day.[10] By the mid-19th century, while scores of specialized journals had sprouted, books and monographs still ruled the roost. Darwin made more of a splash with his commercial 1859 bestseller *On the Origin of Species by Means of Natural Selection* than with the antecedent Darwin-Wallace paper delivered at the nonprofit Linnean Society on July 1, 1858.[11] Things began to change when *Nature* hit the decks in 1869; it has remained a profitable commercial enterprise—and an example for many of us—ever since.

### It Has Not Escaped Our Notice

The mold was struck for the modern scientific paper between the two world wars. Responding to the needs of a self-expanding hierarchy of research universities and professional societies, a standard set in 1927 gained acceptance worldwide.[12] Today the acronymic IMRaD formula (Introduction, Methods, Results and Discussion) is now required by all reputable journals.

But there's always been wiggle room around the canonical IMRaD format; most journals are enlivened by letters to the editor, rebuttals, conference proceedings, abstracts of meetings, news reports, etc. Walter Benjamin's description in 1931 of the marketplace of print still applies to the market in scientific ideas:

> *Today there is hardly a gainfully employed European who could not, in principle, find an opportunity to publish somewhere or other comments on his work, grievances, documentary reports, or that sort of thing. Thus, the distinction between author and public is about to lose its basic character.*[13]

He would have loved texting and Twitter; I can imagine his pleasure at running his thumbs over the passing comments and pertinent grievances as he "follows" and "unfollows" as both author and reader.

In this context, one can only imagine what the epochal Watson-Crick paper would look like these days on *PLoS*. Their 1953 paper was written as a Letter to the Editor in *Nature* and never underwent peer review. John Maddox, editor-in-chief at the time, later admitted that "the Crick and Watson paper could not have been refereed: its correctness is self-evident." That's a matter of dispute, as we'll see.[14] The Watson-Crick paper begins with:

> *We wish to suggest a structure for the salt of deoxyribose nucleic acid (D.N.A.). This structure has novel features that are of considerable biological interest. . . .*

The ending of the paper is of course perhaps the best known in scientific prose:

> *It has not escaped our notice that the specific pairing we have postulated immediately suggests a possible copying mechanism for the genetic material.*

But for many of us, the real action is in the acknowledgments at the end:

> *We are much indebted to Dr. Jerry Donohue for constant advice and criticism, especially on interatomic distances. We have also been stimulated by a knowledge of the general nature of the unpublished experimental results and ideas of Dr. M. H. F. Wilkins, Dr. R. E. Franklin and their co-workers at King's College, London.*[15]

One need only to imagine what tweets, twoops, formal corrections and comments might decorate these passages on *PLoS One* today. Pauling, Chargaff, Avery, Meselson, Cairns, Donohue, Perutz, Franklin and Wilkins would have had their say:

> *This structure has novel features* **COMMENT:** YEAH! HY-DROGEN BONDING, LINUS! *which are of consider-able biological interest.* **COMMENT:** FOR WHICH I WROTE THE CHEMISTRY, ERWIN **FORMAL COR-RECTION:** IT'S THE GENETIC MATERIAL, YOU FOOLS!, GENES! OSWALD
>
> *It has not escaped our notice that the specific pairing* **FORMAL CORRECTION:** BASE PAIRING A/T=G/C, ERWIN *we have postulated immediately suggests a possible copying mechanism for the genetic material.* **COMMENT:** LIKE WHAT? CONSERVED? SEMI? MATT **COM-MENT**: MORTAL OR IMMORTAL? CAIRNS
>
> *We are much indebted to Dr. Jerry Donohue for con-stant advice and criticism, especially on interatomic distances.* **FORMAL CORRECTION:** SEZ YOU! I TOLD YOU ABOUT THE KETO TO ENOL TAUTOMERS. YOU KNEW SQUAT FROM THE CHEMISTRY! JERRY *We have also been stimulated by a knowledge of the general nature* **FORMAL CORRECTION:** I SHOWED YOU THEIR PICTURES, MAX *of the unpublished experimental results and ideas of Dr. M. H. F. Wilkins, Dr. R. E. Franklin* **FOR-MAL CORRECTION:** YOU PEEKED, "DARK LADY" *and their coworkers at King's College, London* **COMMENT:** OUR TWO FOLLOWING PAPERS ARE DATA, YOURS IS A LEAP, MAURY.

### Arcades to the Border

Walter Benjamin (1895–1940), the quintessential European intellect and literary omnivore, would have loved having a COMMENT and FOR-MAL CORRECTION option at his fingertips. His lifelong passion, the *Passagenwerk* or *Arcades Project*, was posthumously published as a handsome volume of more than 1,000 pages.[16] The *Project* is a rough alloy of insights, interruptions and wisdom as dense and fragmented as the imaginary DNA discourse above. The work owes as much to Dante as to Benjamin's fellow refugee from Hitler, Sigmund Freud. Benjamin may have set out to trace the roots of modern Europe to its Oedipal childhood in the 19th century, but in retrospect, the *Arcades Project* spells out some of the reasons that Europe went to hell in the 20th century.

More to the point: much of the *Arcades Project* prefigures the home page of a social network on the web. Benjamin literally explores a *network*: the *linked* indoor shopping arcades of 19th century Paris, the Passages.[16] I imagine a Benjamin today, reincarnated as the perennial *flaneur*, who *follows* a path in the Arcade of Panoramas. He stops occasionally at one *site* or another site. The flaneur ambles (*surfs*) along a protected space (*MySpace*) in which bustling crowds are reflected in shiny *Windows*. He adjusts his cravat in a storefront mirror (*Facebook*), and when the *bell-tone* rings in his pocket, he takes out his timepiece (*Blackberry*). He looks past his mirror image (*YouTube*), to find two generations of *followers* (*Twitter*).

Were Benjamin to log on to Twitter, he'd have thousands of tweets on hand to send to generations of followers. His essay on "The Work of Art in the Age of Mechanical Reproduction"[2] could serve as a source for the ages. The essay describes how advances in the reproduction of visual arts have led to a loss of the expressive "aura" that original paintings, drawings and sculpture have about them, as they are endlessly reproduced and manipulated. Benjamin was ambivalent about the ever increasing power of mechanical reproduction. The medieval woodcut was succeeded by engraving and etching at the end of the Middle Ages. Then came lithography in the 19th century, which next "permitted graphic art for the first time to put its products on the market, not only in large numbers as hitherto, but also in daily changing forms." And then "lithography was surpassed by photography . . . which led to film." Film was the ultimate medium for Benjamin. In the century of the common man, film was art without "aura" and accessible to all:

> *Magician and surgeon compare to painter and cameraman. The painter maintains in his work a natural distance from reality, the cameraman penetrates deeply into its web. . . . Thus, for contemporary man the representation of reality by the film is incomparably more significant than that of the painter, since it offers, precisely because of the thoroughgoing permeation of reality with mechanical equipment, an aspect of reality which is free of all equipment.*[2]

I can see Benjamin now tweeting, now twoopsing, now blogging, now surfing, now scrolling. His thumbs move quickly over the tiny keys—the

sandbox of images in sight. He tweets directly to Biz Jones and the other followers of WB (his *nom-de-tweet*), an upbeat quote from Paul Valéry (1928).
Valéry and WB were sure that other great gadgets would soon supplant
celluloid film:

> *Just as water, gas, and electricity are brought into our houses*
> *from far off to satisfy our needs in response to a minimal effort,*
> *so we shall be supplied with visual or auditory images, which*
> *will appear and disappear at a simple movement of the hand,*
> *hardly more than a sign.*[17]

Pretty good prediction, no? Isn't that "simple movement of the hand"
what the thumbs are doing these days on a Blackberry? The quote is also
about twice the 140 characters that Biz Stone permits, but heck, WB could
have split it in two.

Benjamin might have sent another tweet—one that's carved into his
tombstone overlooking the Spanish border town of Port Bou. After a decade
of exile in France, Benjamin, as Marxist and Jew, had escaped from a Vichy
prison in 1940. In the French border town of Banyuls he was provided with
a map, a guide and a U.S. passport issued by a courageous American consul in Marseille, Hiram Bingham IV. On September 25, 1940, Benjamin
joined a partisan trek on a clandestine, nighttime path over the Pyreanees.
At the Spanish border, however, the Franco police seized him. Benjamin
was an illegal immigrant: he had no French exit visa and Marshall Pétain
had just signed a pact with Hitler that guaranteed return of such refugees.
He was brought to a cheap hotel in Portbou on the Spanish side and told
he would be returned to France on the next day. Fearing deportation, de-
pressed, and alone, Benjamin died during the night, an apparent suicide
from an overdose of morphine. The tweet WB might send is his epitaph:

THERE IS NO DOCUMENT OF A CULTURE,
WHICH IS NOT AT THE SAME TIME
A DOCUMENT OF BARBARISM.[18]

It's less than 140 characters. But, I'd bet that Benjamin would have been
at home in our new world of texting and tweets, blogs and handhelds. In
the Age of Twitter, he'd be ready to play in the sandbox of ideas, or—better
yet—to sit upon a bough to tweet to lords and ladies of Byzantium of what
is past and passing and to come.

# Epigenetics in the Adirondacks

*From so simple a beginning endless forms most beautiful and most wonderful have been, and are being, evolved.*

—Charles Darwin, 1859[1]

*Organized beings are reproduced, from generation to generation, with characters identical to the ones that they possessed during their first emergence.*

—Louis Agassiz, 1869[2]

*If you want the correct explanation/ Why embryos grow into men/ The Alsatian begets an Alsatian/ The hen's egg gives rise to a hen/ Why insects result from pupation/ Why insects grow out of a seed/ Then just murmur epigenetics/ For that is the word that you need.*

—Barnet Woolf, On C. H. Waddington's 50th Birthday, 1955[3]

*It suddenly struck me what an important part fashion had played in the development of our science . . . and it occurred to me that this would make a rather suitable subject for a female president at a Paris congress. So this is what I am going to talk about: fashion in cell biology.*

—Dame Honor B. Fell, "Fashions in Science," 1960[4]

## Epigenetic Fashion

FOR SEVERAL SEASONS NOW, the runways of biological fashion have been awash in epigenetics. We define epigenetics as stable changes in cellular phenotype independent of changes in the Watson-Crick pairing of DNA bases. Since epigenetic changes can be passed to daughter cells in mitosis or meiosis, they can cross generations.[5,6] Open any journal to see the latest epigenetic style. Over 5,000 publications on epigenetics are listed on PubMed in 2011 alone; the field has its own journal—*Epigenetics*, of course (http://www.landesbioscience.com/journals/epigenetics/); and—in keeping with current fashion—a "human epigenome project" has been launched. (http://www.epigenome.org). As *Vogue* might say, epigenetics is so NOW! Experimentally, epigenetics has been used to explain sex-linked inheritance and gene silencing in mammals, type switching in yeast and apparent, if not real, mutations in plants. Clinically, it's been held responsible for cancer, schizophrenia, alcoholism and obesity. Epigenetics may also explain the discordant course of disease in identical twins and is credited for making possible the cloning of Dolly the sheep.[7] No wonder *The Times* (London) headlined its primer on epigenetics: "How your behaviour can change your children's DNA."[8] Indeed, if acquired characteristics can be passed to our progeny, the unthinkable is abroad again: Emma Whitelaw noted that some of "[these new] findings raise the spectre of Lamarckism and epigenetics is now being touted as an explanation for some intergenerational effects in human populations."[6]

But the new epigenetics is not only Lamarck *vs.* Darwin revisited. Our interest in the inheritance of methylated characteristics is probably just the latest swing of the fashion pendulum between nature and nurture in human affairs. Aristotle proposed that humans developed from the interplay of inherited nature, or "preformation," and nurture, which he called "epigenesis."[9] Once genetics became a science, and "evo" (evolution) joined "devo" (development) in university departments, nature and nurture were revived in the guise of Mendelian *vs.* non-Mendelian heredity. The nature/nurture dialectic was also formulated as Weismann's notion of an immortal germ plasm *vs.* mortal protoplasm; and in Ernst Mayr's "soft" inheritance *vs.* the "hard" kind.[10] More recently, Richard Dawkins's selfish gene (nature) has been countered by E. O. Wilson's altruistic biology (nurture).[11,12] Right now the epigenetic style is in high fashion with us and, like all high fashions, attracts the attractive. Dame Honor Fell, my Cambridge mentor, anticipated this sentiment in her presidential address to the First International Congress of Cell Biology in Paris:

> *Sartorially speaking we are probably not an outstandingly fashionable group but, where our research is concerned, we can be as fashion-conscious as the most elegant woman in this City. In science, as in the world of dress,* fashions *recur. . . .*[4]

Research over the last decades has indeed reversed the fashion of Watson-Crick determinism. Under the rubric of epigenetics, Jean-Baptiste Lamarck is back, one chemical link after another. First came DNA methylation of dinucleotides (CpG) in the helix itself, then other heritable modifications of histones and/or nucleosomes were identified: methylation, acetylation, phosphorylation, ubiquitylation, sumolyation *und so weiter*. Genes for micro and small interfering RNA's also undergo epigenetic change in the course of gene silencing. And now, histone codes have joined the "methylomes" of DNA on the catwalks of our journals.[13,14]

## A Landscape at the Cocktail Hour

Not all the fashion critics are in love with the new style. The Nobel Laureate Joshua Lederberg suggested that the terms "nucleic" and "nonnucleic" would better describe the chemical modifications of nature or nurture; he found epigenetics unnecessary for explanations of gene activation.[15] Nor is Mark Ptashne convinced by all those acetylated histones:

> As a glance at the literature will reveal, histone modification—acetylation, phosphorylation, methylation, and so on— are now often explicitly called "epigenetic modifications." This despite the fact that, so far as I am aware, no histone modification has been shown to be heritable.[16]

In fact, the new epigenetics didn't begin with studies of heritability or evolution, it addressed problems of development: "devo" anteced evo. The notion was proposed by the Cambridge polymath Conrad Hal Waddington (1905–1975) while working with Honor B. Fell at the Strangeways Research Laboratory on nuclear-cytoplasmic interactions. Waddington used Fell's hanging-drop cultures to identify "organizing" factors that could dictate the fate of limbs, eyes or spinal column in the chick embryo. In a 1939 treatise, he formulated the nature/nurture problem in the then fashionable terms of nucleus and cytoplasm:

> The interaction of these constituents [nuclear and cytoplasmic] in the egg gives rise to new types of tissue and organ which were not present originally, therefore development must be considered as "epigenetic."[17]

Waddington argued that notions of genotype and phenotype were not adequate or appropriate to describe the difference between an eye and a nose in the course of individual development. But while he was quite innocent of the biochemistry behind the biology of evo/devo, he seems to have anticipated modern signal transduction pathways in a telling analogy:

> *The general picture that emerges from all this is that the embryonic cell is rather like a room where a cocktail party is going on with the radio set with press button tuning in the centre of it. The switching on of a particular battery of genes, controlling the synthesis, say, of nerve proteins, corresponds to pressing one particular button which brings in a programme precisely from one station. But you may succeed in getting this button pressed by jogging the elbow of somebody at the other side of the room, who stumbles against the next man, and so on down the line until somebody finally falls against the radio set, and this may be sufficient to click on whichever of the tuning buttons is most insecure.* [18]

Plus or minus our modern repertoire of acetylases, demethylases, the Ras's, ING's and Runx's, that seems like a pretty good description of epigenetics at a cocktail party. But his studies of embryology soon led him afield to ask why "The Alsatian begets an Alsatian / The hen's egg gives rise to a hen." [3] Waddington contended that classical Darwinian evolution ought to be amended by taking phenotype into account. He formulated a notion called "canalisation" to explain reciprocal interactions between environment and phenotype on the one hand (nurture) and genotype and phenotype on the other (the nature of nature). [18]

To illustrate his point, he drew another analogy, one a little tougher to handle than his cocktail party. He introduced the notion of an "epigenetic landscape" in which a heritable phenotype—a cell, say, or a nucleus—is propelled like a craft on a swift stream flowing between high ridges (the environment). Successive generations of the same phenotype will tend to seek the same path and the phenotype will become fixed, or "canalized" regardless of the variability of its environment or genotype. [19] Alsatians will beget Alsatians, because Alsatian ridges select for survival of those fittest for the Alsace. [3] Waddington's epigenetic landscape proposed a dynamic model of evolution that should pass muster for all seasons of fashion. It's also an extension of the Darwinian origin of species that's worth teaching in these days of "intelligent design."

## The Landscape of Intelligent Design

On a fall weekend, surrounded by the red and gold foliage of the Adirondacks, I was reminded of Waddington's epigenetic landscape: the ridges were high, the streams free-flowing and we stayed overnight in a green valley. We had crossed the very lakes and streams over which a century and a half ago Darwin's archenemy had passed.[20] Harvard's Jean Louis Rodolphe Agassiz (1802–1873) was the preeminent natural philosopher of his day. His stay at what came to be called the Philosophers' Camp at Follensby Pond has become legendary. The scenery, as described by his companion Ralph Waldo Emerson, evokes Waddington's landscape:

> . . . *a small tortuous pass*
> *Winding through grassy shallows in and out,*
> *Two creeping miles of rushes, pads, and sponge . . .*
> *Northward the length of Follansbee we rowed,*
> *Under low mountains, whose unbroken ridge*
> *Ponderous with beechen forest sloped the shore.*
> *A pause and council: then, where near the head*
> *On the east a bay makes inward to the land*
> *Between two rocky arms, we climb the bank . . .*[21]

Agassiz, Emerson, and their colleagues—ten gentlemen and scholars from Boston—had been corralled by the artist-diplomat William J. Stillman to engage in a long canoe trip to camp among the streams, lakes and forests of the then-virgin Adirondacks. Emerson was pleased by his scientific companion:

> *We flee away from cities, but we bring*
> *The best of cities with us, these learned classifiers,*
> *Men knowing what they seek, armed eyes of experts.*
> *We praise the guide, we praise the forest life;*
> *But will we sacrifice our dear-bought lore*
> *Of books and arts and trained experiment,*
> *Or count the Sioux a match for Agassiz?*[21]

The expedition included not only Agassiz and Emerson, who was the country's leading moral philosopher, but also Jeffries Wyman, the eminent Harvard anatomist; the jurist Ebenezer Hoar; James Russell Lowell of *The Atlantic Monthly*; Dr. Oliver Wendell Holmes's brother John and other Brahmin elders. They "swept with oars the Saranac," endured "Hard fare, hard bed and comic misery—the midge, the blue-fly and mosquito," they ate what they shot, fished and plucked by roaring fires—all with the aid of local guides and woodmen. "Ten famous intellectuals, playing Indian" a fan of the area quipped.[22]

Indeed, there could have been more of them. Stillman had asked Henry Wadsworth Longfellow to join the group.

> *"Is it true that Emerson is going to take a gun?" Longfellow asked. "Yes" [Stillman] replied." "Then I shall not go," said Longfellow, "somebody will be shot."* [23]

Under canvas in the Philosophers' Camp, each Boston nabob did what he did best. Lowell shot at an osprey and climbed a pine. Emerson sniffed pine, wrote verse and fired two shots from his rifle. He missed both the bear at which he took aim and his hunting partners. (*Pace*, Mr. Cheney.) Agassiz was the one most completely in his element. With Wyman as assistant, he "dissected the slain deer, weighed the trout's brain," captured "lizard, sala-mander, shrew crab, mice, snail, dragon-fly, minnow and moth." Agassiz regaled his colleagues with stories of his student days with Baron Cuvier in Paris, his theories of glaciers and the Ice Age, and the mating habits of Euro-pean fish. He preached opposition to the "classifications" of Carl Linnaeus and Ernst Haeckel and pleased his companions by insisting that Lamarck and Darwin were wrong. Species did not arise from others, they were all created in infinite variety by—you guessed it:

> *In the infinite variety of their structure, I see the direct action*
> *of an intelligence, made manifest in the most diverse ways,*
> *rather than the development of successive generations by means*
> *impossible to define.* [2]

It is difficult these days to understand why bluff Agassiz, whose legacy to modern science is dubious, was so highly treasured by the lettered folk of his time. Lowell, Holmes, Wyman and Emerson—indeed all the lights of Boston—were in awe of this Swiss polymath. But if we measure his works against those of the grand system-builders of Europe—Linnaeus, Buffon, Lamarck, Cuvier, Charles Lyell, Darwin—or look into the accounts of his shady business ventures, it becomes easier to understand why he left Swit-zerland for fame and fortune in the New World.

Having left his first wife to die of tuberculosis in Europe, Agassiz insin-uated himself into the Brahmin aristocracy. Indeed, his second wife, Eliza-beth Cabot Cary Agassiz, was to become the first president of Radcliffe, and her paternal grandfather was Col. Thomas H. Perkins, the benefactor who established the Perkins Institution for Dr. Samuel Gridley Howe, husband of Julia Ward Howe of "The Battle Hymn of the Republic." In an 1863 letter to Howe, appointed by Lincoln to head the American Freedmen's Inquiry Commission, Agassiz expressed his views of fixed racial types:

> *Conceive for a moment the difference it would make in future ages for the prospect of republican institutions and our civilization generally, if instead of the manly population descended from cognate nations, the United States should hereafter be inhabited by the effeminate progeny of mixed races, half Indian, half Negro, sprinkled with white blood. In whatever proportion the amalgamation may take place, I shudder at the consequences.*[24]

He also found time to explain to Dr. Howe that the coloreds were genetically inferior, that mulattos were sure to prove sterile and that amalgamation of the races was biologically unsound.

Agassiz was convinced that the purpose of God is fixed and could be read in the facts of nature. "Study Nature, Not Books" reads the motto of Agassiz, now preserved in the library of the Marine Biological Laboratory at Woods Hole. "Study Books, You'll Learn More" cracked Eugene Bell, the late and wise embryologist. Haeckel would have helped the Harvard professor.

## Thought's Newfound Path

In camp at Follensby, evolutionary thought was interrupted by electric news. Visitors arriving by canoe reported that the first trans-Atlantic submarine cable between Newfoundland and Ireland had been successfully connected. The philosophers exulted over the news:

> *Of the wire-cable laid beneath the sea,*
> *    and landed on our coast, and pulsating*
> *With ductile fire. Loud, exulting cries*
> *From boat to boat, and to the echoes round,*
> *Greet the glad miracle. Thought's new-found path*
> *Shall supplement henceforth all trodden ways,*
> *Match God's equator with a zone of art,*
> *And lift man's public action to a height*
> *Worthy the enormous cloud of witnesses.*[21]

Emerson was energized by the possibility of communication between the continents. He imagined that it might someday be feasible to use electrical impulses to mimic—or to substitute for—the machinery of human mentation. Electricity for Emerson was:

> *A spasm throbbing through the pedestals*
> *Of Alp and Andes, isle and continent,*
> *Urging astonished Chaos with a thrill*
> *To be a brain, or serve the brain of man.*[21]

But even the Adirondacks, that Arcadia of lake and that epigenetic landscape of gorge and canalization, contained the serpent of preformation. Emerson had learned from Agassiz that if races were created all at once by the direct action of intelligent design (*l'action immédiate d'une intelligence se manifestant*), why Nature must be destiny, and Self-Reliance its gospel. How appropriate, then that "a manly population descended from cognate nations" had found thought's newfound path. How happy we should be that:

> *It is not Iroquois or cannibals,*
> *But ever the free race with the front sublime,*
> *And these instructed by their wisest too,*
> *Who do the feat, and lift humanity.*[21]

These days we've relegated preformation to the history books. We're busy working out the latest fashions in signal transduction on the web and thought's newfound paths make their wireless way to every nook of the Adirondacks. There is a new problem, however: the signal-to-noise ratio is getting harder to overcome. Dame Honor had already issued that warning in Paris in 1960:

> *Our scientific world is becoming like a crowded cocktail party,*
> *in which everyone shouts a little louder in the hope of making*
> *himself heard, until at last the volume of speech is such that*
> *almost nothing can be distinguished.*[4]

One hopes that cacophony in science may itself be a transient fashion, which, like the biological fancies of the Philosophers' Camp, will also pass with time. Alas, the serpent of biological racism remains abroad in the land, well tended by Darwin-deniers. On the other hand, those heirs of Agassiz are so very not NOW.

# A Nobel Is Out of Order:
# "J.Lo" vs. Hypatia of Alexandria

*Robert Edwards wins 2010 Nobel Prize in Medicine for In-Vitro Fertilization.*

—*The Washington Post,* October 4, 2010[1]

*Without Edwards, there would not be a market on which millions of ovocytes are sold. . . . In the best of cases they are transferred into a uterus but most probably they will end up abandoned or dead, which is a problem for which the Nobel Prize winner is responsible.*

—Vatican spokesman Carrasco de Paula, October 5, 2010[2]

*I Would Never Have In Vitro. . . . When it comes to family and relationships, I'm quite traditional. Just because of the way I was raised. And I also believe in God and I have a lot of faith in that, so I just felt like you don't mess with things like that.*

—Jennifer Lopez, Interview with *Elle,* 2010[3]

*As Hypatia comes forth to her academy, she is assaulted by [Saint] Cyril's mob—an Alexandrian mob of many monks. Amid the fearful yelling of these barelegged and black-cowled fiends she is dragged from her chariot, and in the public street stripped naked. In her mortal terror she is hauled into an adjacent church, and in that sacred edifice is killed by the club of Peter the Reader. They outraged the naked corpse, dismembered it, and, incredible to be said, finished their infernal crime by scraping the flesh from the bones with oyster-shells, and casting the remnants into the fire. . . . The leaden mace of bigotry had struck and shivered the exquisitely tempered steel of Greek philosophy.*

—John William Draper (1864), *The Death of Hypatia,* 415 AD[4]

## A Trifecta for the Nobels

IN 2010, TWO ANNOUNCEMENTS OVERLAPPED in the news. In late September, pop media broke the story that Jennifer Lopez had been appointed judge for the 20th season of *American Idol,* perhaps the highest-rated TV show in the history of television.[5] A week later, the winners were announced of the 110th round of the Nobel Prizes, the highest-rated awards in the history of science, letters and good works.[1,6,7] Cheery delight greeted the news of J.Lo's ascent to the throne of pop glory: "I'll take my cues from you from now on" a fellow judge told the Bronx-born "triple threat actress."[3] Responses to the new Nobel laureates were far less sanguine: the cries of dissent set a record.

While in years past one or another of the Prizes, chiefly those for Literature and Peace, have aroused controversy, never have three of the awards met with such diverse displeasure. In 2010, a powerful American senator censured the Prize for Economics; the Chinese rulers waxed wroth over the Prize for Peace; and the Vatican lashed out at the Prize for Medicine and/or Physiology. To offend Washington, Peking and Rome, the Scandinavians must have done something right!

Commenting on the award of the Nobel Prize in Economics to Peter Diamond of MIT, Republican Richard Shelby of Alabama maintained that "while the Nobel Prize for Economics is a significant recognition, the Royal Swedish Academy of Sciences does not determine who is qualified to serve on the Board of Governors of the Federal Reserve System."[6] The senator— best known to biologists as a staunch opponent of stem cell research—had single-handedly blocked Diamond's appointment to the governing board of the Fed. He complained that the new laureate had an insufficient knowledge of macroeconomics. Reminded that Peter Diamond was a former teacher of the Fed's present chairman, Ben Bernanke, the senator insisted that Professor Diamond "would be learning on the job."

The Chinese government exceeded Senator Shelby in displays of pique. After the Nobel Peace Prize was awarded to a jailed dissident, Liu Xiaobo, the Peking Ministry of Foreign Affairs complained that "The Nobel Committee has completely violated the goals of, and has desecrated, the Peace Prize."[7] China's official news agency blasted the dissident, accusing him of wanton eagerness "to ingratiate himself with and kowtow to his Western masters. . . . The Nobel Peace Prize is a political 'reward' tossed to him by his Western masters." The Chinese canceled several cabinet-level meetings with the Oslo government and placed the laureate's wife under house arrest.[8]

Displeasure in Rome completed the trifecta. The Vatican-based International Federation of Catholic Medical Associations was "dismayed" at the award of the Nobel Prize in Physiology and/or Medicine to Robert Edwards for in-vitro fertilization.[9] The head of the Pontifical Academy for Life, Ignacio Carrasco de Paula, was even more acerb. He blamed Edwards for creating a market in embryos and failing to protect human life. The Vatican spokesman proclaimed, "I find the choice of Robert Edwards completely out of order."[2]

The Prize may be out of order, but it laid the foundations for modern reproductive biology. Many considered it long overdue. The first test-tube baby of 1978, Louise Brown, gave birth to her own daughter in 2009 and over four million other babies are alive today thanks to in-vitro fertilization. Edward's collaborator, Patrick Steptoe, the doctor whose technique of laparoscopy made IVF possible, died in 1988 and was ineligible for the prize. Edwards himself is 85 and will have to join this year's Peace Prize recipient on the list of "Nobel No-Shows."[10] Liu is in jail and Edwards is frail.

Be the Prize late or not, the discovery remains as seminal—in both senses—as when Edwards and Steptoe made it:

> *I'll never yet forget the day I looked down the microscope and saw something funny in the cultures. I looked down the microscope* [c1965] *and what I saw was a human blastocyst gazing up at me. I thought: "We've done it!"*[11]

## The Back-Up Plan and Infanticide

We can chalk up the long lag-time between the invention of IVF and that "out-of-order" Prize to systems of belief that antedate any microscope. In reference to her film *The Back-Up Plan*,[12] Jennifer Lopez expressed a common attitude toward assisted reproduction of any kind: "IVF isn't the way these things are supposed to happen," the actress told *Elle*, "deep down I really felt like either this is not going to happen for me or it is. You know what I mean? And if it is, it will. And if it's not, it's not going to."[3] Right on J.Lo! Opposition to IVF is based on faith.

Joseph Goldstein, chairman of the Lasker jury, presented Robert Edwards with a 2001 Lasker award, comparing his contribution to Darwin's:

> *As one might expect, when we challenge our conception of humanity, we arouse controversy. Indeed,* The Origin of Species *and the first test-tube baby ignited two of the most violent*

*controversies in the history of biology and medicine. If the hu-
man species evolved by natural selection instead of by Divine
creation, then the Bible cannot be literally true. If human beings can be conceived in test tubes by scientists, then the act of
conception has lost much of its mystery.*[13]

Like Darwin's magnum opus, the work of Edwards and Steptoe shook
the fans of natural law. While the Catholic Church became the chief opponent of IVF and related techniques of assisted reproduction, other objections—both lay and clerical—also delayed the path to public acceptance.

In 1971, seven years before Louise Brown was born, bioethicist Leon
R. Kass—later a strong opponent of human embryonic stem cell research—
worried over risks to mother, child and social values. Kass proclaimed in the
*New England Journal of Medicine* that "One cannot ethically choose for a
child the unknown hazards that he must face, and simultaneously choose
to give him life in which to face them."[14] In the same year, DNA's James
Watson, warned Edwards that "You can only go ahead with your work if
you accept the necessity of infanticide."[15] The recent flaps over the ethics
and politics of human embryonic stem cell research—which relies in part
on the use of embryos discarded in the course of IVF—have added other
recruits to the ranks of zygote defense.

An opponent both of IVF and stem cell research, Richard Land of the
Southern Baptist Convention, told Reuters that "To argue that one human
being is more developed and therefore in greater need in no way justifies the
cannibalizing of another to benefit him."[16] A director of a Catholic bioethics institute in Britain, a group strongly opposed to human stem cell research of any sort, chimed in with the complaint that IVF "has led directly
to the deliberate destruction of millions of human embryos."[16] Indeed. In
fact, millions of children are alive *because* of Edwards and Steptoe.

Spokesmen of other major religions, Jews, Muslims and Buddhists, have
urged "limits on test tube babies" and linked IVF to sacrilege.[17,18] But the
most articulate, well-reasoned opposition to IVF and related techniques has
been mounted by the Catholic Church. Beginning with Pope Paul VI's encyclical letter of July 1968, *Humanae Vitae,* the Church has insisted on natural
means of reproduction, therefore disallowing contraception, surrogate motherhood and the like.[19] This position was reinforced by the Church document
*Donum Vitae* (1987) signed by then Cardinal Joseph Ratzinger and approved
by Pope John Paul II. It urged regulation of new reproductive technology such
as "test-tube" fertilization, surrogate motherhood and artificial insemination

because of "possible unforeseeable and damaging consequences for civil society."[20] By 2008, the position was made more explicit. *Dignitas Personae* ruled that most forms of artificial fertilization were to be excluded on the grounds that they replaced the conjugal act as a means of reproduction.[21]

## Darwin and St. Cyril

The Vatican's consistent support of God-ordained, traditional reproduction in the face of every modern development in reproductive science has been linked to denial of Darwinian evolution.[22]

It's a straight line from the first popes—Pius IX and Leo XIII—who were faced with the "unproven theory" that man descended from monkey. In 1883 Leo XIII issued edicts that insisted on the literal interpretation of Adam as the only, single antecedent of man; and that a virgin gave birth. He then responded to the challenge of Victorian science with an appeal to the faithful in *Adiutricem* "because, if ever there was a time when love and veneration of the Blessed Virgin [must be] awakened, it is in these days so bitterly anti-religious."[23] The Church authority to whom Leo appealed in support of this encyclical was St. Cyril of Alexandria (ca. 376–444), the chap who expelled the Jews from Alexandria, dismantled the Alexandrian library, established a theocracy and "brought down the mace of bigotry" on poor Hypatia in 415.[4,24]

In the fall of 2010, while naysayers were completing their anti-Nobel trifecta, other mischief was afoot. Bewitched politicians twittered that "There is not enough evidence [for evolution] to make it a fact [because] the world began as the Bible in Genesis says, that God created the Earth in six days, six 24-hour periods. And there is just as much, if not more, evidence supporting that."[25] If—as the saying goes—history repeats itself, the first time as tragedy, the second time as farce, we might remember what happened the first time fundamentalist bigots took over a tolerant, diverse, polity: Alexandria.

## Hypatia

Alexandria in the fifth century was a world hub of commerce and learning, that in the words of John William Draper, historian and Darwinist,

> . . . *could vie with the city of Constantinople itself. The city was full of noble edifices—the palace, the exchange, the Caesareum, the halls of justice. Among the temples, those of Pan and Neptune were conspicuous. The visitor passed countless*

*theatres, churches, temples, synagogues. . . . On a flight of a hundred marble steps stood the grand portico with its columns, its chequered corridor leading round a roofless hall, the adjoining porches of which contained the library, and from the midst of its area arose a lofty pillar visible afar off at sea.*[4]

Hypatia (ca. 355–415) occupied a post of honor in the academy of the museum and vast library. According to her student Synesius of Cyrene, she taught mathematica and philosophy: natural and theoretical, exact and Platonic. Hailed as the "first woman scientist," she invented the hydrometer and an early version of the astrolabe. Hypatia's legend is long; Voltaire and Edward Gibbon celebrated her tale as a tableau of Greek reason and eclipsed by the dark centuries to come.[26,27] Draper lists her among the first of the martyrs of science—along with Michael Servetus, Giordano Bruno and Galileo.[4] She died at the hands of Cyril's mob, both for what she represented and for supporting the secular ruler of the town, Orestes, in his opposition to the expulsion of the Jews by Cyril. A contemporary described the event:

*Cyril, accompanied by an immense crowd of people, going to their synagogues—for so they call their houses of worship— took them away from them, and drove the Jews out of the city, permitting the multitude to plunder their goods. Thus the Jews, who had inhabited the city from the time of Alexander the Macedonian were expelled from it, stripped of all they possessed and dispersed some in one direction and some in another.*[28]

Hypatia dissented, and her story—as told by Socrates Scholasticus— continues.[28] Hypatia was held in honor because of "the self-possession and ease of manner, which she had acquired in consequence of the cultivation of her mind, she not infrequently appeared in public in presence of the magistrates." The only woman at the academy, she was not shy when addressing an audience of men, who, because of her extraordinary "dignity and virtue admired her the more." Yet even she fell a victim to political jealousy. "It was calumniously reported among the Christian populace" that it was Hypatia who prevented Orestes from coming to terms with Cyril. Cyril's mob, "carried away by a fierce and bigoted zeal, whose ringleader was a reader named Peter, waylaid her returning home, and dragging her from her carriage, they took her to the church called Caesareum." The gruesome story was finished by Draper.[4]

It may seem farcical these days to compare martyrdom in Alexandria to today's faith-based objections to the facts of Darwin or those of Edwards and Steptoe. But there's a pertinent quotation attributed to Hypatia:

> *Fables should be taught as fables, myths as myths, and miracles as poetic fancies. To teach superstitions as truths is a most terrible thing. The child mind accepts and believes them, and only through great pain and perhaps tragedy can he be in after years relieved of them. In fact men will fight for a superstition quite as quickly as for a living truth—often more so.*[27]

No matter how innocent, sentiments like "So I just felt like you don't mess with things like that" remind one of what happens when the child mind accepts and believes. There's a difference between pop culture and the facts of science: the bottom line is that the world is round, humans evolved from an extinct species and Elvis is dead.

# Epigenetics and Alma Mahler

*The principal matter of importance in this is an artificial coloring, probably with India ink, through which the black coloring of the skin in the region carrying the stripes is said to have been faked. Therefore it would be a matter of deception that presumably will be laid to me only. Who beside myself had any interest in perpetrating such falsifications can only be very dimly suspected. . . . I hope that I shall gather together enough courage and strength to put an end of my wrecked life to-morrow.*

—Paul Kammerer, *Science*, 1926[1]

*Rather than committing fraud, it seems that Kammerer had the misfortune of stumbling upon non-Mendelian inheritance at a time in which Mendelian genetics itself was just becoming well accepted.*

—A. O. Vargas, "Did Paul Kammerer Discover Epigenetic Inheritance?" 2009[2]

*I kept records, very exact records. That, too, irritated Kammerer. Somewhat less accurate records with positive results would have pleased him more.*

—Alma Mahler-Werfel, *Mein Leben* [My Life], 1960[3]

## The Origin of the Specious

WHEN A SCIENTIST COMMITS SUICIDE a few weeks after a paper in *Nature* shows that he faked a critical experiment, we might assume the case closed. However, if the scientist is the Viennese biologist Paul Kammerer (1880–1926) and if the case is the notorious "Case of the Midwife Toad," then the docket remains open. Kammerer's wacky campaign to prove the heritability of acquired characteristics is as topical today as on September 23,

1926, when he shot himself on an Austrian hill. Not one of his controversial "discoveries" has ever been duplicated, his notion that musical talent is heritable remains on the shelf and his contention that the Prohibition laws in America would induce a genetically superior race of teetotalers is too absurd to consider.[4–6] However, his star has risen once again, thanks to a bold claim that he is the father of epigenetics[7] and a new look at his place in the history of image manipulation.[4]

Kammerer's claim in 1909 that male midwife toads pass on acquired nuptial pads to the fourth generation of their progeny has made him a hero to armchair generals in the nature/nurture wars.[8] His career was first resurrected by a great writer, Arthur Koestler, who pleaded Kammerer's innocence in "The Case of the Midwife Toad."[9] Reviewing that book, Stephen Jay Gould thought that Kammerer's tale of the toad was probably all right; he cut Kammerer slack for his progressive politics and his "penetrating intelligence."[10] That was 1972, and the case became moot for a generation as DNA became RNA, etc., etc. Suddenly, in October 2010, A. O. Vargas of Santiago, Chile, re-reinterpreted the midwife toads in the light of modern epigenetics and paternal imprinting. Based on his analysis of "parent-of-origin" data buried in Kammerer's toad papers, Vargas argued that Kammerer was "the actual discoverer of epigenetic inheritance."[2,7] Editorial fanfares in the *Journal of Experimental Zoology* and *Science* praised Vargas for unearthing what "may have been the first demonstration of a recently recognized [*sic*] phenomenon: epigenetics."[11,12] Shucks, and I thought Waddington charted the epigenetic landscape, Shirley Tilghman got imprinting right and Jean-Pierre Changeux planted the flag for epigenesis in synaptic affinity.[13–15]

## Alma Tell Us . . .

Well, epigenetics may be looking for a father, but I'm persuaded that Kammerer's short, frantic career was based on error and—his word in the suicide note—deception. Indeed, error and deception pop up in his science, his personal life and his public statements. It's rare for a scientist to commit suicide after his work has been refuted. It's rarer still for the suicide note to be published in *Science*.[1] However, it's completely unheard-of for folks in the same lab to snitch on their colleague in print. Franz Megusar, who was Kammerer's coworker in Vienna's *Biologischen Versuchsanstalt* (the Experimental Biology Station of Vienna, also called the Vivarium), included these comments in the annual *Proceedings of the Conference of German Scientists* in 1913:

*The processes that Kammerer reports I could not confirm, nei-
ther in his experiments nor in my own even though I have been
monitoring his imprecisely executed experiments constantly for
nearly ten years. . . . Kammerer's representations contain crude
untruths and falsifications of the actual circumstances.*[16]

Megusar was not the only close observer of Kammerer's sharp practice
at the Vivarium. There was Alma Mahler, the femme fatale of 20th-century
Vienna modernism. Married, successively, to Gustav Mahler (the sympho-
nies), Walter Gropius (the Bauhaus) and Franz Werfel (*The Song of Berna-
dette*), her close intimates included Gustav Klimt and Oskar Kokoschka
(the painters) and Gerhart Hauptmann (the playwright), as well as a legion
of others.[17] She was celebrated on both sides of the Atlantic, winding up in
a ditty of Tom Lehrer's:

*Her lovers were many and varied,
From the day she began her—beguine.
There were three famous ones whom she married,
And God knows how many between. Alma, tell us!
All modern women are jealous.
Which of your magical wands
Got you Gustav and Walter and Franz?*[18]

One of those "many between" was Paul Kammerer, known to the *gra-
tin* of Vienna as a womanizer, a "wizard of lizards," an amateur musician
and friend of Gustav's. Although Alma later confessed that the biologist
was not up to his artistic competitors—"he was the clown of my whole
circle"—they had a brief fling. He had more of a fling than she: in one of his
passionate letters, Kammerer threatened Alma that he would shoot himself
over Gustav Mahler's grave if she did not marry him. She didn't; Walter was
groping about. She was soon on to the next but not before warning Kammerer's current wife (he had begun his own beguine) to "get that pistol out
of the house."[17] Yet, for two years, 1911–1912, between Gustav and Walter, Alma assisted Kammerer at the Vivarium. She helped Kammerer with
spotted salamanders, the skin patterns of which were alleged to be heritably
affected by the color of sand on which they were reared. Her autobiography
describes the reptile work and its later refutation by E. G. Boulenger of
the London Zoo. It corroborates Megusar's description of Kammerer as a
mendacious observer:

*He wanted positive results in his research so much that he would unconsciously depart from the truth. This trait explains to me his later problems when English researchers showed that "on further examination, his [salamander] experiments proved invalid." On that occasion the mimicry of salamanders was the subject. These experiments, with which I helped, were rushed into print and not accurately documented.*[3]

The salamander experiments that Alma had witnessed established Kammerer as the father of photographic image manipulation.[19,20] Kammerer explained to his protesting editor that he had inked in colored spots on photos of experimental salamanders, as "The glare from the skin gave the impression of spots where none were present and the spots that were present were washed out in the glare."[21] The passage is worthy of Danny Kaye in *The Court Jester*—"The pellet with the poison's in the vessel with the pestle, the chalice from the palace has the brew that is true"—but image manipulation is serious business these days.

Although there is general agreement over what may legitimately be "image-enhanced" in biological images,[22] matters of deception pop up all too often. Picassos of Photoshop and Raphaels of the Raster map have been involved in a flock of "Retractions" or "Editorial Expressions of Concern" in recent pages of *Science, Proceedings of the National Academy of Sciences USA* and the *FASEB Journal*.[23–27] The mischief has been followed by extensive self-flagellation and almost universal calls for forensic examination of each pre-publication image. Using such programs to detect artful dodgers, our colleagues at the *Journal of Cell Biology* report that an astounding 25 percent of the articles accepted for publication contain at least one image that violates the journal's guidelines.[28] What hath Kammerer wrought?

### Black Pads and Spurious Siphons

Our Viennese friend was finally undone, not by image manipulation but by India ink. Kammerer, it seems, had mastered another art form: the manipulation of actual specimens. We know now that his unrepeatable salamander data were the result of fraud or foolishness but so was the midwife toad. Kammerer admitted in his suicide note that he had found his preserved salamanders "blackened," ditto the pads of the midwife toad![12]

Kammerer had claimed that land-dwelling, male midwife toads, which carry their partner's fertilized eggs on their own back, would not only acquire nuptial pads when arid conditions forced them into water but also that these black, horny protrusions could be passed on to male progeny to the F6 generation. The inheritance of that acquired characteristic would permit land-dwelling toads to take on the phenotype of their water-dwelling cousins. (Male frogs and toads that live in ponds develop nuptial pads with which to clasp their slippery partners while mating.) G. K. Noble, a herpetologist from the American Museum of Natural History, went to Vienna to examine Kammerer's last pickled specimen with "inherited" nuptial pads and discovered that India ink had been used to create the illusion of the blackened structure.[29] Noble's 1926 article in *Nature*, which detailed "this matter of deception," led to that shot in the head.

Kammerer's last words about the matter pointed to another origin of the specious: "Who besides myself had any interest in perpetrating such falsifications can only be very dimly suspected."[1] Thomas Hunt Morgan of Columbia wrote to his friend G. K. Noble shortly after Kammerer's suicide note appeared in *Science*:

> *Kammerer has done one more dirty trick in trying to put the fraud over on to one of his assistants. Remember that this is not the first time, either, that he has been caught and all responsible people will, I think, draw the same conclusion.*[30]

It was certainly *not* the first time that he had been caught. Kammerer claimed that the critical experiment, which "proves the inheritance of acquired characteristics," was the heritability of hyper-regenerating siphons in the tunicate, *Ciona intestinalis*.[31] Kammerer purported to show that when the two siphon ends of this protochordate were amputated, the new siphon tubes became longer upon regeneration than the original tubes and that the complete elongation was inherited by the next generation. The whole experiment supposedly involved two sequential siphon regenerations, after which a regeneration of a lower section of the body, containing the gonads, was caused to occur before the animals were crossed for the next generation.[32]

Three refutations of these claims have been published. In 1923, Harold Munro Fox of Cambridge wrote in *Nature* that "I have repeated the amputation experiments and find that the regenerated siphons do *not* grow beyond the normal length."[33] In 1930, a Russian scientist, Julius Wermel, published 26 pages of data about *Ciona* regeneration and concluded: no elongation, no heredity.[34] The most damning refutation was that of J. R. Whittaker, a former director of the Marine Biological Laboratory (MBL) at Woods Hole, who spent two summers following up Arthur Koestler's charge that someone, somewhere try to repeat the *Ciona* work.[35] Whittaker had followed Kammerer's directions to the letter and concluded:

> Ciona *siphons did not, of course, regenerate longer after their surgical removal. But most telling was the gonadal regeneration part of Kammerer's supposed experiment. This involved a completely lethal operation from which animals could not recover. . . . I was left with no remaining doubt that the* Ciona *results were also an invention of Paul Kammerer's high-strung imagination.*[32]

## A High-Strung Imagination

The conclusion Whittaker drew in 1985 could apply to Kammerer's entire output. Kammerer had presented evidence for his neo-Lamarckian notions about the inheritance of acquired characteristics to two audiences. To persuade biologists, he had performed all of those experiments—refuted or not—and published scores of technical papers and seven full-length books. For the wider public of his day, his lecture demonstrations, books and pamphlets attracted broad attention. To the crowds who came to listen in Europe and America, he must have sounded like E. O. Wilson on one day and Deepak Chopra on the other. In 1912, he'd written a short work assuring the world that musical talent was heritable; he dedicated the volume to Alma and her daughter.[5] Jacques Loeb, writing from the MBL, commented on the notion:

> Kammerer *. . . claims that an interest in music on the part of parents produces offspring with musical talent. In such claims much depends upon the subjective interpretation of the observer. The writer is not aware that there is at present on record a single adequate proof of the heredity of an acquired character.*[36]

However, lack of adequate proof didn't bother Kammerer. On a visit to America, he gave a series of lecture demonstrations, collected as *The Inheritance of Acquired Characteristics* (1924).[37] Scrutinized today, it's a collection of Ripley-derived Believe-It-or-Nots. The volume features his drawing of a Japanese dancing mouse with a "mutilated" tail and two of her young progeny also "born with mutilated tails," an observation on the heritability of acquired characteristics that might have surprised readers whose forefathers had long practiced the art of circumcision. He also hit the press with this headline in the *The New York Times:*[6]

### BIOLOGIST TO TELL HOW SPECIES ALTER
### DR KAMMERER, "DARWINS SUCCESSOR"
### ARRIVES FROM VIENNA FOR LECTURES

A scientist who did not hesitate to explain to his editor that he was forced to manipulate spots on photos of salamanders had no trouble assuring the *Times* that "The next generation of Americans will be born without any desire for liquor if the prohibition law is continued and strictly enforced."[6] He also assured those packed audiences at his lectures that Germany and Austria were far ahead of the United States in their effort to improve the race by cultivating physical fitness and eugenical breeding in accord with his personal notion of *Korperkultur und Rasse* (Bodily Fitness and Race Culture).[38]

I'm afraid that Kammerer's story remains pertinent today, when our journals print retractions of articles that have sported manipulated images, duplicated data and fabricated authorship. Somebody desperately wants them to be okay! We live, these days, with virtual reality and biased avatars; it's hard to pick out fact from faction. Fraud tends to be ignored by those who agree with the conclusion it reaches, whether facts support it or not. For a lie to persist, or to be resurrected like the midwife toad, there has to be an audience that requires belief. But, as Robert Graves observed in both the social and natural sciences,

> *Theft is theft and raid is raid*
> *Though reciprocally made.*

# Inflammation Is Complicated:
# From Metchnikoff to Meryl Streep

*In* It's Complicated *Ms. Meyers transforms a divorced couple into a romantic couple, which suggests a belief in love enduring even after a marriage dies. That sounds wonderfully romantic or a prescription for pathology, maybe both. [But] she wants, as her title implies, to complicate that formula. We have met the enemy, Ms. Meyers seems to be suggesting, and she is firmer—and younger.*

—Manohla Dargis, *The New York Times*, Dec. 25, 2009[1]

*But infection also has its counter. The attacked organism defends itself against the little aggressor. It protects itself by all the means at its disposal to destroy the invader. The organism digests substances introduced outside the digestive tract by means of an inflammatory response.*

—Elie Metchnikoff, 1905[2]

*The end result [of inflammation] [often] is not defense; it is an agitated committee-directed, harum-scarum effort to make war, with results that are remarkably like those sometimes-observed in human affairs. . . .*

—Lewis Thomas, 1971[3]

## The Cardinal Signs on Film

ONE DOESN'T USUALLY COME AWAY from a Hollywood flick with thoughts of Metchnikoff and inflammation. On the other hand, *It's Complicated* features each of the cardinal signs of inflammation: redness and swelling with heat and pain, leading to loss of function. It also shows an attacked organism defending itself against a little aggressor. The film is set in Boniva/Viagra-land, in which

44

a divorced couple (Meryl Streep and Alec Baldwin) catches up with each other after 15 years apart. Streep's in *pain*: their three grown kids have left the nest empty, and Baldwin's married to a *hot*, young chick. The couple meets up to celebrate their son's college graduation. They lose composure at a bibulous *function*. Streep turns *red*, and Baldwin prances. Sure enough, they wind up in bed. Back home, Streep's circle *swells*: the kids return with significant others, the firm, young chick has a lovable kid of her own, and then there's Steve Martin, to whom Streep is loosely attached. Agitated, harum-scarum efforts (à la Thomas) are launched; action and reaction follow. But, at last, the empty nest becomes a love nest when Streep and Martin exit laughing as the final credits roll. The end of this so-so film pays tribute to a hot, new area of experimental biology: the resolution of inflammation.[4]

The literature is awash in inflammation these days (Table 1). I'm sure that we apply the term much too loosely. Every condition in which someone's humor goes awry, or a cell waxes wroth is chalked up, willy-nilly, to "inflammation."

A glance at Table 1 shows that inflammation has been held responsible for aging and obesity, depression and cancer, not to speak of osteoporo-

**Table 1:** *Conditions Associated with "Inflammation" in Papers Listed by PubMed**

| Infection | 135,713 |
|---|---|
| Cancer | 33,602 |
| Rheumatoid arthritis | 9,166 |
| Atherosclerosis | 8,411 |
| Obesity | 4,401 |
| Aging | 3,682 |
| Depression | 2,293 |
| Alzheimer's | 1,231 |
| Osteoporosis | 791 |
| Schizophrenia | 194 |
| Fibromyalgia | 175 |
| Erectile dysfunction | 82 |
| Anorexia nervosa | 41 |
| Hirsutism | 28 |

For "Inflammation" as of 9/15/2010.

* Of 332,445 listed.

sis and erectile dysfunction. It's complicated—too complicated, perhaps. Most of these papers simply describe that one or another cell has made or responded to one or another *mediator* of inflammation. So perhaps we'd better go back to the original definition of inflammation and to its original cause: the battle between a host and an army of microbes.

Redness and swelling with heat and pain—*rubor et tumor cum calore et dolore*—have been recognized as the four cardinal signs of inflammation since the writings of Aulus Cornelius Celsus (ca. 25 B.C.–ca. 50 A.D.). The fifth—*functio laesa* (loss of function)—was added by Rudolf Virchow in 1858.[5] Doyens of the field agree that although one can have one or another of these signs without inflammation, it ain't inflammation until one has at least four.[6] Emotions, such as turning crimson like Streep ogling Baldwin or pining in pain like a flummoxed Steve Martin, are not inflammation. But, if you've caught a streptococcus in your throat or been stung by a wasp on your eyelid, only *then* are you inflamed: you've seen redness and swelling with heat and pain firsthand. As for *functio laesa*, call your doctor.

We've learned recently that our brigades of evil humors (now called ILs, IFNs, and inflammasomes) respond to levels of cellular concern that wouldn't raise a small pimple on the chin.[7,8] So, let's watch our language and not cry inflammation whenever one or another cytokine cries for attention from its array on the chip. On the other hand, what if Metchnikoff's little aggressors, the microbes, really *are* responsible for conditions that rank near the top of the list in Table 1? Isn't human papilloma virus infection related to cancer?[9] Hasn't rheumatoid arthritis been linked to gum infection by a microbe horribly called porphyromonas gingivalis?[10] Aren't *Chlamydia pneumoniae* found in the our aging aortas?[11] It's complicated.

## The Battle Hymn of Microbes

It may be a stretch to extract classic pathology from the plot of a Hollywood film, but not the other way around. For me, the classic accounts of inflammation spin a tale more artful than any recent film of Streep's. In its simple, concise expression, Celsus's *rubor et tumor cum calore et dolore* owes much to artful poet Horace, whose *Nullius addictus iurare in verba magistri* (I am not bound over to swear allegiance to any master) became the motto of the Royal Society.[12] Celsus, who is sometimes confused with Anders Celsius of thermometry (1701–1744), was describing the typical reaction of flesh to microbes, not, as nowadays, the increase of a mediator or cytokine of some type. Indeed, evolutionary biology teaches that although inflammation may help the *individual* cope with cuts and bruises, it was primarily designed to protect the *species* from lethal epidemics.

By 1905, when Robert Koch was awarded his Nobel Prize for spotting the agents of anthrax and tuberculosis, it became evident that inflammation served to defend us against microbes, not miasmas. Three years later, Nobel Prizes were awarded to Paul Ehrlich for his work on humoral immunity (antibodies) and to Elie Metchnikoff for his work on cellular immunity (phagocytosis). Their work established that the body uses both strategies to identify and destroy microbial invaders. In the battle against infection, we call our losses "infection" and our victories "immunity."

The discourse that announced Metchnikoff's discovery was that of Darwinian survival:

> *When the aggressor in this struggle is much smaller than its adversary the result is that the former introduces itself into the body of the latter and destroys it by means of infection. . . . But infection also has its counter. The attacked organism defends itself against the little aggressor. It protects itself by interposing a resistant membrane, or it uses all the means at its disposal to destroy the invader.*[2]

The major "means at its disposal" was the arousal of phagocytes (or "eating cell," which was Metchnikoff's term). The phagocyte engulfs the invaders in a resistant membrane (now, a phagolysosome) and digests them within the vacuole by means of cytases (lysosomal enzymes). When these enzymes are mistakenly released into tissues, inflammation results.[13] Many of us remain convinced that this mechanism lies at the root (literally) of inflammation in several common conditions, but again, it's complicated.

Following Metchnikoff's lead, it's been common to describe inflammation in terms of 19th-century warfare. "Struggle," "aggressor," "invader," etc., are not only the language of Darwinian survival but also, the rhetoric of imperial conquest. The distinguished pathologist Joseph McFarland summed up the field of inflammation from Rudolf Virchow to the advent of antibiotics:

> *To many, the situation here encountered resembles a battlefield on which the leukocytes meet the invading bacteria and contest their further increase and invasion until they triumph, and, the infection overcome, the inflammation subsides.*[14]

The microbe hunters drew their images of battle from romantic accounts of skirmishes waged more often than not by splendidly equipped British or French troops against primitively armed "lesser breeds." Appropriately, Anatole France remarked that *"le principe fondamental de toute guerre coloniale est que l'Européen soit supérieur au peuple qu'il combat."* (The fundamental principle of colonial war is that Europeans are superior to the nations they fight.)[15]

Metchnikoff, Ehrlich, Julius Cohnheim and J. G. Adami peered through the lenses of microscopes to watch battalions of white cells engage insurgent microbes. Caught in the imperial zeitgeist, Metchnikoff drew the lesson of Darwin's struggle for survival from the evidence of phagocytosis:

> *The diapedesis of the white corpuscles, their migration through the vessel wall . . . is one of the principal means of defense possessed by an animal. As soon as the infective agents have penetrated into the body, a whole army of white corpuscles proceeds towards the menaced spot, there entering into a struggle with the micro organisms. The leukocytes, having arrived at the spot where the intruders are found, seize them after the manner of the Amoeba and within their bodies subject them to intracellular digestion.*[2]

Metchnikoff was not the only champion of white corpuscles; we owe their classification as basophils, eosinophils and neutrophils to Ehrlich's doctoral thesis about aniline dyes. Blue, red and white colors depended on the way in which positive or negative charges lined up in the granules of the phagocytes; aniline dyes marked the combatants as clearly as red or blue tunics identified cavalry units in the Franco-Prussian War of 1870. Indeed, aniline dyes were the response of German synthetic chemistry to French control of the colonies that yielded natural dyestuff. Ehrlich's studies with dyes, which introduced the language of colors and fixation, helped Germany to become the arsenal of chemotherapy and had the happy side effect of launching immunochemistry. Ehrlich elaborated the dictum that *Corpora non agunt nisi fixat.* (Bodies do not act unless fixed[16]).

This principle was the basis not only for the first treatment of syphilis—Salvarsan—but also for a modern ligand-receptor theory. It's been argued that Ehrlich's doctrine also applies to the means whereby cells (white cells and platelets) stick to vessel walls and to each other in the course of inflammation; we call those processes hetero- and homotypic cell adhesion, respectively. We now recognize that these processes are tightly controlled by signals launched via pattern recognition molecules, present on vessel walls and on circulating cells.[7] Ehrlich's dictum can now be re-phrased as *Cellulare non agunt nisi fixata.*

## Domestic Disputes: It's Complicated

Paul Ehrlich and his contemporaries were sure that the body had a *horror-autotoxicus*—an incapacity to turn its defensive weapons into tools of self-destruction. But, time and events have overturned that conviction. In the 1950s, Roitt and Doniach, among others, proposed the notion of autoimmunity: inflammation in the thyroid was caused by antibodies directed against a self-antigen.[17] Horror, indeed! Since then, we learned that the tools of acquired immunity as well as those of innate immunity could be self-destructive. Counterinsurgency can lead to self-injury. Divorce hurts, as Streep explains in the film.

We have unearthed an array of unlimited weapons for warfare against the invisible armies of microbes, and each of them can be turned back on our own tissues. Our cells in combat are bathed by evil humors with acronyms worthy of the Pentagon—C5a, C3a, IFN, IL, MIF, etc.—to which they can react in both pro- and anti-inflammatory modes. Signals from "helper" and "suppressor" lymphocytes trigger cellular signals via JAKs and the STATs, the NF-kBs and so forth and so on. The CIA would indeed have trouble deciphering the battle orders. Sadly, the collateral damage we inflict is often to ourselves.[7,8,18]

Nowadays, when we speak of inflammation, we often speak of un-planned mischief, as Lewis Thomas predicted:

> *I suspect that the host is caught up in mistaken, inappropri-ate, and unquestionably self-destructive mechanisms by the very multiplicity of defenses available to him, defenses which do not seem to have been designed to operate in net coordi-nation with each other. . . . If, to push the analogy, there were no limit to the number of people who could set off for northern Minnesota at the season of the great fly-over of geese and no limit on the type and power of the weapons to be used by each, we would undoubtedly observe, what with M-16's, howitzers, SAM missiles, lasers, and perhaps tactical nuclear rockets, considerably more destruction of people than geese, of host than invader.*[3]

It's complicated.

# An *Arrowsmith* for the NASDAQ Era:
## *Extraordinary Measures*

*Mary was a child of seven or eight. Martin found her lips and finger-*
*tips blue, but in her face no flush. . . . He thanked the god of science*
*for antitoxin and for the gas motor. It was, he decided, a Race with*
*Death "I'm going to do it—going to pull it off and save that poor kid"*
*he rejoiced.*

—Sinclair Lewis, *Arrowsmith*, 1925[1]

*Genzyme Corp. won approval from the Food and Drug Administra-*
*tion to sell the first treatment for Pompe disease, a rare and devastating*
*enzyme-deficiency disorder that often causes extreme muscle damage*
*and heart failure. The new treatment, called Myozyme®, will be ex-*
*traordinarily expensive, likely costing around $200,000 a year. Gen-*
*zyme, based in Cambridge, Mass., says it spent $500 million over eight*
*years to develop the treatment.*

—*The Wall Street Journal*, April 29, 2006[2]

*John took a deep breath, looked around the room, and lifted his right*
*hand to a switch and pressed the "on" button. He saw the switch open*
*and the clear liquid [Myozyme®] begin dripping . . . Megan's heart*
*monitor began to beep and . . . soon everyone was laughing through*
*their tears.*

—Geeta Anand, *The Cure*, 2006[3]

### If I Don't Write a Good Business Plan, My Child Will Die[4]

IT'S NOT OFTEN THAT YOU SEE scientists in movies point to "glycobiology"
or "phosphorylated glycans" on a blackboard, unless they're fabricating ava-

tars to destroy the world. But, along comes *Extraordinary Measures,* a film in which Indiana Jones—I mean Harrison Ford—puts down his hat and bullwhip to become a lovable curmudgeon and "another beloved American archetype: the renegade scientist."[5] Reminds one of that first archetype of an American movie scientist, Martin Arrowsmith. However, *Extraordinary Measures* isn't about one lovable scientist—curmudgeon or not—finding a magic medicine in one "Eureka!" moment. If you want that story, you'd better go back to Martin Arrowsmith and his bacteriophage. In his day, bench moved to bedside without venture capital.

In *Extraordinary Measures,* the dialogue may come from the playbook of storage diseases,[6] but the story is cast as a NASDAQ release. The movie begins with two cute kids afflicted with Pompe's disease, a glycogen storage disease that cripples their muscles and heart; it ends with a cure dripping into their veins. The movie doesn't tell us a lot about the science of Pompe's disease: it certainly doesn't describe the efforts of Rochelle Hirschhorn or Arnold Reuser[7] to clone the gene for acid maltase, the enzyme that's missing or mutated in Pompe's. We don't hear about Elizabeth Neufeld and Roscoe Brady,[8,9] who traced how enzymes can get into lysosomes in the first place and how acid maltase can be replaced. And we're not treated to the story told by Christian de Duve in his 1974 Nobel Prize lecture:

> *What we did not suspect in the beginning was that the failure of lysosomal enzymes to act at their normal site could also cause serious diseases. This fact was brought home to us in a rather surprising fashion by Géry Hers, who in 1962 diagnosed glycogen storage disease type II as being due to a severe deficiency of a lysosomal enzyme.* [10]

*Extraordinary Measures* is based on *The Wall Street Journal* reporter Geeta Anand's book, *The Cure: How a Father Raised $100 Million—and Bucked the Medical Establishment—in a Quest to Save His Children.* [3] Plus or minus some tear-jerking moments, the film is as true to Anand's book as a popular film can get, although the book is as close to science as a financial journalist can get. Both tell us that the road to a cure is paved with venture capital and that in this game, one hundred million bucks is Jacks-to-Open. Ford plays a fictional physician/scientist named Robert Stonehill, roughly modeled on William Canfield, a glycobiologist, formerly of the University of Oklahoma. In the movie, Stonehill is tracked down to a fictional lab in the boondocks by John Crowley, a go-getting entrepreneur.[11] Actually,

Crowley did not "buck the medical establishment"; in real-life, Crowley met Canfield at an NIH meeting in 1998, called to include every figure in the Pompe's disease medical "establishment."[12]

Crowley, a pharma executive, then with Bristol-Myers Squibb, was out to develop, at any cost and with any effort, the newest treatment for his two children who were afflicted with Pompe's disease. At that NIH meeting, he learned that Canfield had formulated a novel polyphosphorylated glycan adduct of the acid maltase. Crowley bought the notion that Canfield's enzyme might gain far easier entry into the glycogen-stuffed lysosomes of Pompe's disease than any other. But setting up animal models of Pompe's disease was expensive; new labs and equipment were needed. A new biotech company was the answer, and Novazyme became its name.

### Don't Try This at Home

*Extraordinary Measures* presents a realistically filmed pageant of any start-up biotech, the ups and downs of compound after compound synthesized and tested in the dish, the dog-and-pony shows to lure the first investors, the tense presentations before "mezzanine-round" venture capitalists. In the movie, as in real life, Novazyme succeeded. Based entirely on preclinical data, Canfield and Crowley persuaded a large, top-ten biotech company, Genzyme of Cambridge, Massachusetts, to buy out Novazyme for something north of $137 million. Canfield told his local paper, "It was one of the largest biotech transactions in 2001. It was an unusually high value for a pre-clinic [*sic*] company. We had been in operation for only 27 months. This wasn't typical; do not try this at home."[13]

In the event, as with many such acquisitions, it's by no means clear whether *The Cure* led to the cure. Certainly the drug that flowed into the child's vein in the film, and into hundreds of real-life children, was not Canfield's enzyme. Genzyme had set four independent groups the task of finding which of four enzyme preps would work best in several preclinical models. Only one of these was the Novazyme version. By 2002, after reviewing the results of "The Mother of All Experiments," Genzyme was ready to go into clinical trials with the winner of the contest. It was not *The Cure* but the real cure, Myozyme®, a formulation developed internally at Genzyme. Almost a decade later, it's hard to tell whether the Canfield/Crowley effort spurred or delayed the development of Myozyme®. Buying Novazyme was costly, but the move forestalled any future competition with Genzyme's other lysosomal enzyme products.[14]

## Pompe's Gift

And now for the kicker: When Myozyme® was approved by the FDA, Henri Termeer, the Dutch CEO of Genzyme, was proud to announce that the drug was developed, not only with scientists at Duke University (Yuan-Tsong Chen, Priya Kishnani et al.) but especially with A. J. J. Reuser, A. T. Van der Ploeg et al., of Erasmus Medical Center in Rotterdam.[15] Rotterdam, indeed!

Joannes Cassianus Pompe (1901–1945) would have been pleased at the contribution of his native land to the treatment of a disease that he was the first to describe. Pompe, a native of Utrecht, received his MD from the University of Amsterdam and after a short stint in pathology at Nijmegen, returned to Amsterdam in 1939, where he moved up the academic ladder. He had presented his doctoral thesis on "*Cardiomegalia glycogenica* (glycogenic cardiomegaly)" in 1936,[16] but perhaps because of the rarity of the condition—and of peace in Europe—there was no follow-up. After the German occupation of Holland in 1940, Pompe became an active member of the Dutch resistance.[17] Eventually, the Germans found a secret radio transmitter in his pathology laboratory and imprisoned him on February 25, 1945. As bad luck would have it, other members of the resistance blew up a railroad line in nearby St. Pancras. In keeping with Nazi practice, he and 19 other prisoners—uncharged and untried—were shot in retaliation. [*Mit 19 weiteren Personen erschossen bei einer Vergeltungsmaßnahme nach dem Anschlag auf eine Bahnlinie. Zwei Wochen später wurden die Niederlande befreit*].[18] Pompe was executed two short weeks before Holland was liberated. Sixty years later, thanks to science and NASDAQ, a cure for Pompe's disease was dripping into the veins of children in the Netherlands.

The moral of *Extraordinary Measures* was foretold by Henry James in *The Golden Bowl*: "For what was science but the absence of prejudice backed by the presence of money?"[19] The moral we can draw from Pompe's biography is that a single observation, by a single scientist, can have a profound impact. It's a gift, so to speak, that's ours for the taking.

## A Nobel Prize Film

Nowhere is the story of the single scientist making a single discovery better told than in the 1925 novel and 1931 film *Arrowsmith* by Sinclair Lewis, who won the Nobel Prize in literature in 1930. Years ago, every medical student in the country was attracted to it; lately, a good number of my students have been led to read *Arrowsmith*, after someone told them that the name of the rock group "Aerosmith" was taken from a "doctor book." They've found it quaint, dated and totally inspiring. The character of Martin Arrowsmith was prompted by the "Microbe Hunters the late Paul de Kruif wrote about: Koch, Pasteur and Ehrlich." Lewis has Martin resolve that if he had to be "a small town doctor he would be such a small town doctor as Robert Koch."[1] However, unlike Aerosmith's grungy millionaires, Martin leaves riches, not for rags but for science; his picaresque career blends the biographies of Paul de Kruif and Lewis's father: research in a bacteriology lab in Michigan, a small-town doctor's life in Wisconsin, basic experimental work at the Rockefeller Institute, and the temptations of money and the flesh. Martin works at the fictional McGurk Institute on bacterial reproduction with a Dr. Gottlieb (Jacques Loeb) and has the Eureka! moment.

Martin's great discovery, set in the 1920s, is the "X Principle," bacteriophage, in real life discovered by Twort and d'Herelle[20,21] at the time of the Great War.

> *"I have observed a principle, which I shall temporarily call the X Principle, in pus from a staphylococcus infection, which checks the growth of several strains of staphylococcus, and which dissolves the staphylococci from the pus in question."*[22]

In due course, Martin finds that various bacteriophages can kill all sort of bacteria, staph, strep, even the *Yersinia* of plague. He rushes to inform the smarmy director of the McGurk Institute, a character modeled after Simon Flexner, that he has a discovery that will kill bacteria without harming the host.

Paul de Kruif, unacknowledged coauthor of *Arrowsmith*, had been dismissed by Flexner in 1923 from the Rockefeller Institute for falsified claims and publicity seeking. De Kruif had written a trendy exposé about the medical establishment for *The Century* magazine under the anonymous signature of K, M. D. But de Kruif was more than a self-destructive iconoclast. As a young researcher, he had performed experiments at the University of Michigan with Frederick G. Novy, which might have put him in a position to discover that DNA was the "transforming principle" of the rough-to-smooth transformation of pneumococcus, a line of investigation that eventually proved crucial to the flowering of DNA.[23] However, after Flexner put a stop to de Kruif's career at Rockefeller,[24] revenge reared its head in the blustering caricature of Dr. Tubbs. Both Novy and Flexner had been on the U.S. Plague Commission that worked to combat outbreaks of plague in San Francisco (1900) and New Orleans (1914–1917).[25,26] Lewis, following de Kruif, has Tubbs exort to Arrowsmith:

> *"I've been thinking, Arrowsmith," said Tubbs. . . . "As I understand it, you've been going along with what Dr. Gottlieb would call 'fundamental research.' I think it may now be time for you to use phage in practical healing. I want you to experiment with phage in pneumonia, plague, perhaps typhoid, and when your experiments get going . . . let's really cure somebody . . . Go cure the plague!"*

In the event, Martin and his wife Leora (played in the film by Ronald Coleman and Helen Hayes) travel to a Caribbean island to stop an outbreak of the bubonic plague by means of a specific bacteriophage that he has developed. He will also do a controlled trial, à la Gottlieb: half will get the phage; the other half, not. On that plague-ridden island, Martin's Swedish colleague is killed by the epidemic that they have been fighting; his death is the heroic death of a Microbe Hunter. Leora is with Martin in the place referred to as Black Water, but when he goes to another island after a heated argument with the Governor General, he implores his wife to stay where she is, believing that it is for the best. One perceives the telltale cigarette on which *Yersinia* have been spilled; Leora has a cut finger, and she is infected. Too late with his phage, Martin finds that Leora has died, forsaken and alone. In 1931, *The New York Times* commented on the film:

> The Screen: A Nobel Prize Novel: *Mr. Colman [is] Martin Arrowsmith, the zealous young doctor and hopeful scientist . . . combating the bubonic plague in the West Indies. Life and death and Arrowsmith's devotion to science are emphasized toward the end, the last scene being depicted in a most dignified way.*[27]

## Phage Redux

Despondent over his wife's death, Martin decides that the controlled trial he has planned would be inhumane; everyone on the island he can reach will be treated! Film and book end with Martin leaving the Institute and the big city to work at a small lab in the country. That's where he would have needed a John Crowley to find him and NASDAQ to bring phage back to life.

The lack of controlled trials for phage therapy has bedeviled clinical and experimental work with bacteriophage.[28] However, phage is crawling back as a therapy, as Joshua Lederberg predicted in 1996.[29] The website Phage International lists 14 biotechnology companies exploiting bacteriophage for the treatments of topical and systemic infections,[30] but injected phages are rapidly cleared by the lysosomal apparatus in liver and spleen,[31] and therefore, claims for such treatments have always seemed spurious. Happily, intravenous preparations have recently been formulated that can escape clearance by our scavenger cells, and topical applications already abound.[32]

Since 1915, phage treatment has wanted only "the absence of prejudice backed by the presence of money." Perhaps the next version of *Arrowsmith* will feature the CEO of a well-funded NASDAQ company (George Clooney in blazer and chinos?) announcing FDA approval of a bacteriophage treatment for one or another plague.

# Sarah Palin and Marie-Antoinette: Post-Traumatic Tress Disorder

*Marie-Antoinette syndrome designates the condition in which scalp hair suddenly turns white. The name alludes to the unhappy Queen Marie-Antoinette of France (1755–1793), whose hair allegedly turned white the night before her last walk to the guillotine during the French Revolution.*

— A. A. Navarini, et al., *Arch Dermatol*, 2009[1]

*Well, that didn't take long. Just 44 days into the job, and President Obama is going gray.*

—*The New York Times*, March 5, 2009[2]

*Friends worried that she appeared anxious and underweight. Her hair had thinned to the point where she needed emergency help from her hairdresser and close friend, Jessica Steele.*

—*The New York Times*, July 13, 2009[3]

*The queen's hair became a site of her attempt to assert personal agency, which could assume negative social and political ramifications, since the assertion of this agency was frequently considered to be in opposition to her responsibilities as queen.*

—Desmond Hosford, *"The Queen's Hair: Marie-Antoinette, Politics, and DNA,"* 2002[4]

## The Hair Care Reform Debate

THE SUMMER OF 2009 WAS MARKED by a political war over the future of medical practice in the United States. Campaigns were launched in print and electronic media, battles flared at a myriad of blogs, tweets and You-Tubes; trench warfare erupted at town meetings between gun-toting citizens and their elected representatives. This dog-day controversy, added to other vexations such as two festering wars and an ataxic economy, took a toll on our political leaders. Both sides of the health care reform (*aka* social justice) debate showed signs of stress at the follicular level. Close observers of President Obama's scalp line noted the earliest warning signal: "44 Days in the White House, and the Hair? Grayer Already" headlined *The New York Times* in March 2009. Presidential advisor David Axelrod was of the opinion that the assumption of high office leads to sudden changes in hair color: "The gray seemed to be on him from the moment he took the oath."[2]

Resignation from office also appears to affect follicular stress. Shortly after Sarah Palin's "Long March to a Short-Notice Resignation" as Alaska's commander in chief, Palin watchers worried that her hair was thinning noticeably.[3] While an occasional psycho-dermatologist attributes stress-induced hair loss to "telogen effluvium,"[5] Palin's Wasilla hairdresser disagrees, insisting that the governor's hair hadn't thinned so much as it had gone blah. She attributed Palin's sudden lackluster coiffure to a stressful "combination of traveling and just being down there in the lower 48."[6] This novel etiology of virtual alopecia was disputed by Maureen Dowd, who dubbed Sarah Palin "a Nixon with hair extensions," that is, the omission of such extensions might appear as thinning hair to the uninitiated.[7] Palin fans quickly rose to her defense against a "liberal smear campaign," praising the beauty and luster of her coiffure.[6]

Indeed, tales abound of the famous and infamous who have suddenly grown gray, or white, or lost their hair altogether, in response to life-threatening stress. Examples of this narrative can be found in the Bible, Shakespeare, Carlyle, Byron and Sainte-Beuve.[8] But perhaps the most celebrated of follicular turnovers is that of Queen Marie-Antoinette of France, condemned to death by a Revolutionary Tribunal on October 16, 1793. She is the eponymic queen of anecdotal dermatology; her syndrome "designates the condition in which scalp hair suddenly turns white."[1] Attention to abrupt changes in a ruler's hair may be a sign of regime change, whether in the court of Louis XVI or Anchorage, Alaska. But, there is really no convincing evidence for the overnight whitening of one's hair; it is probably a lack of pigment of the imagination.[9]

## The Queen's Hair Under Stress

The notion that hair can suddenly whiten overnight is certainly not supported by the actual history of Marie-Antoinette. The queen's hair had attracted careful attention in her lifetime: a scholar notes that

> *The queen's coiffures not only mirrored the evolution of her character, but were sometimes perceived as tools for a dangerous foreigner seeking to undermine the stability of the nation.*[4]

In 1776, for example, a year of national and personal depression in France, courtiers noted that the queen's hair had become short and thinned, setting a new fashion, the *coiffure* à *l'enfant.*[10] But her elaborate, adult tresses returned, and by the time Marie-Antoinette met her untimely end, witnesses had described three separate, overnight bleachings of the royal coiffure. The first was in 1791, after an unsuccessful escape effort, the "Flight to Varennes" from Paris. Disguised as the maid and butler of a Russian baroness (a part played her son's governess), the king and queen attempted a nighttime escape from taunting mobs outside the Tuileries. The royal family almost made it north to Montmédy, near what was then the Austrian Netherlands, only to be captured in the town of Varennes by a crowd who recognized the face of the monarch from his image on coins. Brought back to Paris, the humiliated couple was detained in the Tuileries and later removed to less luxurious confinement at the prison of the Temple. Three days after that night in Varennes, her friend, the Princesse Lamballe noted that the stressful episode had turned the queen's hair "as white as a woman of 70."[11] The queen was 36 years old.

Things continued downhill. Confined in the Temple, separated from her family, she received news of the September massacres of royals and counterrevolutionaries. When the severed head of the Princesse Lamballe was paraded on a pike outside her window, the queen fainted at the sight. She fainted once again when she received news of the king's public execution on January 21, 1793; for a second time, her hair was reportedly turned white by the shock.[12]

She was herself indicted on October 14, 1793. Branded as an agent of her native Austria, as a libertine, and accused of incest with her son, she faced what Sarah Palin could justifiably call a "death panel." As described by Thomas Carlyle:

> *The once brightest of Queens, now tarnished, defaced, forsaken,
> stands here at Fouquier-Tinville's Judgement-bar; answering for
> her life . . . Marie-Antoinette, in this her utter abandonment,
> and hour of extreme need, is not wanting to herself, the imperial
> woman. Her answers are prompt, clear, often of Laconic brevity;
> resolution, which has grown contemptuous without ceasing to
> be dignified, veils itself in calm words. . . . Scandalous Hébert
> has borne his testimony as to many things: as to one thing, con-
> cerning Marie-Antoinette and her little Son,—wherewith Hu-
> man Speech had better not further be soiled. She has answered
> Hébert; a Juryman begs to observe that she has not answered as
> to this. "I have not answered," she exclaims with noble emotion,
> "because Nature refuses to answer such a charge brought against
> a Mother, I appeal to all the Mothers that are here."[13]*

The sentence was pronounced after two days and nights of testimony and a few minutes of deliberation: Death within 24 hours! Her hair,
cropped in preparation for the guillotine, was sparse and white: the third
"sudden whitening" in two years. Antonia Fraser has written that after the
futile flight to Varennes, the abandoned queen had become thin, malnour-
ished and subject to copious uterine hemorrhages; fibroids, other uterine
tumors, or tuberculosis have been postulated.[14]

Carlyle finishes the story:

> *On reaching the Place de la Révolution, [Marie-Antoinette]
> mounted the Scaffold with courage enough; at a quarter past
> Twelve, her head fell; the Executioner showed it to the people,
> amid universal long-continued cries of Vive la République.[13]*

## Oxidative—Not Political—Stress Grays Our Hair

If Marie-Antoinette didn't suffer from her eponymic syndrome, and a
critical look at other supposed examples of sudden hair whitening provides
no clear examples,[8] what is the effect of psychological stress on the hair?
Half a century ago, the eminent dermatologist William Montagna wrote:

> *Locked within the metamorphosing hair follicles in the [hu-
> man] scalp are all the secrets of growth and differentiation.
> When we know these answers, we shall have the key, not to
> hair growth alone, but to all growth, which is, after all, the
> basis of all biological phenomena.[15]*

Since then, we've learned a lot about those follicles on our scalp, each a small homunculus with its own supply of stem cells. These reside in the bulge area, a contiguous part of outer root sheath at the bottom of the follicles.[16] Each follicle undergoes 10 to 30 reproductive cycles in its lifetime. The anagen (active hair growth) phase lasts from two to eight years, the catagen (regression) phase lasts four to six weeks, and the telogen (resting) phase lasts two to three months. The pigmented hair shaft is produced only during anagen, while release of dead hair, the exogen phase, comes at the end of telogen. This cycle requires bouts of melanocyte proliferation (early anagen), differentiation (mid to late anagen) and melanocyte death via apoptosis (during early catagen).[17]

Thus, each hair cycle regenerates a new intact follicular unit, at least for the first ten cycles or so. But soon gray and white hairs appear, explained by what has been called the "free radical theory of graying."[18] Melanocytes make hydrogen peroxide, which is broken down by catalase. But aging follicles fail to generate sufficient catalase, and the excess hydrogen peroxide bleaches the hair shaft directly. Additionally, peroxide not only blocks the active site of tyrosinase, the key enzyme of pigment production, but also destroys enzymes (MSRA and B) that repair peroxide-induced damage to tyrosinase.[19]

Does psychosocial stress produce gray hair? Not a shred of evidence for this notion has been reported in the scientific[8] or popular literature.[20] Our hair becomes gray from using the same chemical used by those who bleach their hair from bottles; as we age we become peroxide grays.

## The Rush Limbaugh Experiment

Is there evidence that psychic stress can cause hair loss? There is, indeed, evidence for a "brain–hair follicle axis" that can influence the hair cycle in experimental animals.[21] Hair follicle cells make their own marijuana-like ligands, endocanabinoids—and respond to them; they also have receptors for many neurohumors, especially substance P.[21-23] Moreover, exogenous psychosocial stresses have been shown to affect the murine hair cycle, as Petra Clara Arck, Ralf Paus and their collaborators have shown. They have developed a model for telogen effluvium—the malady associated with Sarah Palin—in experimental animals. In this model, telogen arrest and hair loss are clearly mediated by substance P.[23]

The experimental methodology deserves note. Mice were stripped of all the hair on their backs from neck to tail by means of a wax/rosin mixture, to set all the new hair follicles into the same hair growth cycle (late anagen). Then they were exposed to an "ultrasound stress" for 24 hours starting on day 14. The source of the ultrasound was a rodent-repellent device set at a frequency of 300 Hz in intervals of 15 seconds. "The stress device was placed into the mouse cage so that the mice could not escape the sound perception." Sure enough, the follicles underwent telogen arrest; the effect was blocked when substance P was inactive. The human equivalent of this sort of psychosocial stress might be reproduced by locking an experimental subject into a telephone booth for 24 hours with a continuous podcast of Rush Limbaugh's voice blasting in the booth. Exposing a governor of Alaska to the psychosocial stress of "traveling and just being down there in the lower 48" wouldn't seem to fit that bill.

Conclusion: Neither the assumption of, nor resignation from, executive office seems sufficient to account for an acceleration of normal human aging.

# Coca-Cola and H. G. Wells:
# Dietary Supplements as Subprime Drugs

*Falcons' defensive tackle Grady Jackson, appealing a four-game suspension from the National Football League (NFL), is suing Balanced Health Products, the maker of StarCaps for "false advertising and unfair business practices."*

—*Atlanta Journal-Constitution*, 2008[1]

*The U.S. Food and Drug Administration (FDA) announced the expansion of the recall of StarCaps-brand dietary supplement. StarCaps is sold in 30-capsule plastic bottles by retail stores nationwide and on the Internet to help "metabolize protein, eliminate bloat, and detoxify your system," as well as to help in weight reduction.*

—UPI, 2008[2]

*More and more products tainted with prescription drugs, including drugs for erectile dysfunction, diabetes, and obesity, are finding their way into the U.S. marketplace. Many are labeled as dietary supplements, however, unfortunately, it is not possible for the FDA to test and identify all tainted products.*

—U.S. Food and Drug Administration, 2009[3]

*"But look here aunt," I said, "It's a quack medicine. It's trash." "There's no law against selling quack medicine that I know of," said my aunt.*

—H. G. Wells, 1909[4]

## The Coke in Coca-Cola

IF OUR ECONOMY TANKED because of subprime mortgages, perhaps the time has come to look at subprime drugs. (I'd call drugs "subprime" if they affect bodily functions without having undergone tests of safety and efficacy by the FDA.) Unfortunately, the $24 billion/year "dietary supplement" industry peddles subprime concoctions that can only be recalled *after* someone blows a whistle. Case in point: a highly publicized lawsuit filed by the Falcons' Grady Jackson preceded a manufacturer's recall of StarCaps slimming capsules and manufacturer recalls of more than 60 other dietary supplements. Among the covert ingredients in these over-the-counter nostrums was bumetanide, a potent diuretic.[5]

This was not the first time a court in Atlanta has heard accusations of hidden ingredients in a patent medicinal. In 1902, an Atlanta courtroom heard evidence that a medicinal "stimulant," Coca-Cola, was liberally laced with caffeine and undetermined amounts of cocaine. The ensuing uproar over secret components, especially cocaine, in patent medicines led to passage of the Pure Food and Drug Act of 1906, provisions of which still guide the FDA. Follow-up litigation ensured that today's Coke no longer contains coke.[6]

The 1902 trial in Atlanta made it clear that Coca-Cola's claims of medical efficacy were inseparable from its addictive content of caffeine and cocaine. Trans-Atlantic publicity following the trial produced a major work of fiction in England—H. G. Wells's *Tono-Bungay*—in which Coca-Cola was recast as the title elixir. Coca-Cola had reached the United Kingdom by 1900, peddled as a medicinal pick-me-up, a "scientific combination of coca leaves and kola nuts" that "cured headaches, calmed the nerves, strengthened the muscles and renewed the vigor of the intellect."[6] The Tono-Bungay of Wells' novel is a stimulating brew of secret ingredients, one of which is so intoxicating that it "Cocks their tails!"[7] The elixir is peddled by a snake-oil millionaire, Edward Ponderevo, who sells it "at an unconscionable price by methods that would shame a confidence man."[8] The character is a dead ringer for Asa G. Candler, "God's Capitalist," who began by concocting Botanic Blood Balm and went on to build a personal empire in Georgia by making Coca-Cola a household name.[9] Wells's narrator, George, is a young lad brought into his uncle's patent medicine empire, who, after rising to the top of the snake-oil world, admits that the whole pharmacopeia of tonics such as Tono-Bungay was "unmitigated fraud by any honest standard, the giving of nothing coated in advertisements for money."[10]

Wells had begun his career writing a textbook of biology but made his fame and fortune as a man of letters (*The Time Machine*, *The War of the Worlds*, etc.). After writing *Tono-Bungay*, a dark, sci-fi masterpiece worth reading by anyone today, Wells was hailed by William James as the Tolstoy of the English world.[11] Wells, as a sometime Fabian socialist, was also a severe critic of laissez-faire and took swipes at Britain's lack of commercial regulation.

As *Tono-Bungay* teaches, there was no law against quack medicine at the dawn of the 20th century, either in the United States or in Britain. Yet, by and by, muckraking novels such as *Tono-Bungay* and Upton Sinclair's *The Jungle* (1906) lit the fires of reform in Edwardian England and Teddy Roosevelt's America. On both sides of the Atlantic, the Progressive Era saw the pendulum swing to law and order from greed and laissez-faire. But not in Atlanta: Asa Candler, the Coke magnate who led the fight against income taxes, excess-profit taxes and cocaine labels, also "clashed with the progressive trends toward government-imposed standards for the manufacture of his product" and was elected mayor of Atlanta in 1917.[9] That was then, this is now; our family is awash in diet Coke.

### Papaya, Valerian, and Garlic

I was reminded of the Coca-Cola trial and *Tono-Bungay* by news of Grady Jackson's class-action suit for "false advertising and unfair business practices." The suit was filed not only against the Beverly Hills socialite founder of the company, Nikki Haskell, but also against stores that sold these dietary supplements, such as the Vitamin Shoppe and General Nutrition Center (GNC). Money passed hands all around: At $160/month, StarCaps had star appeal: "Kathie Lee Gifford was enthusiastic about them on the *Today* show. Retailers like GNC and the Vitamin Shoppe sold them, no prescription required."[5]

Alas, the hefty stars of professional football had followed Hollywood's over-the-counter advice. In the 2007–2008 football season, a half-dozen pros (of the Tennessee Titans, New Orleans Saints and Minnesota Vikings) also failed their urinary steroid tests and faced four-game suspensions by the NFL. The behemoths blamed StarCaps for lacing its product with bumetanide, a loop diuretic banned by the NFL. Football players know that bumetanide will yield false-positives in urinary steroid tests, and it doesn't take a physiologist to know that it produces weight loss by the bucket.[3,5]

There is no mention of bumetanide in the disingenuous list of StarCaps ingredients. The capsules are said to contain a "mix of the digestive enzyme in the papaya and the diuretic qualities in the garlic [that] result in weight loss for users. The capsules also include valerian, an herb that is native to Europe and Western Asia. Today it is used often for its properties as an herbal stress reliever, but valerian was used by early herbalists for its diuretic properties as well."[12] However, omission of bumetanide isn't the only flaw in the description of the capsules. StarCaps ads also failed to note that papain is an effective protease that can be absorbed topically, that valerian has proven soporific effects in humans and that the magnesium salts contained in the capsule affect the gut.[13–15]

The lawsuits filed by the football players caught the attention of the FDA, which since 1994 has had no effective regulatory authority over dietary supplements, unless these nostrums cause harm or are doped with prescription drugs.[3] Within a month of Jackson's suit, the manufacturer recalled StarCaps, which was followed by expanded manufacturer recalls of more than 60 other weight-loss supplements found to contain sibutramine (a centrally acting, serotonin-norepinephrine reuptake inhibitor), rimonabant (a cannabinoid receptor antagonist), phenytoin (better known as dilantin) and phenolphthalein (a pH indicator and laxative, banned in the United States as a carcinogen). The manufacturer recalls of those supplements were accompanied by the following FDA advisory:

> *Many of these products are marketed as dietary supplements. Unfortunately, FDA cannot test and identify all weight loss products on the market that have potentially harmful contaminants in order to assure their safety. Enforcement actions and consumer advisories for unapproved products only cover a small fraction of the potentially hazardous weight loss products marketed to consumers on the Internet and at some retail establishments.*[3]

That's a "small fraction" of those $24 billion spent on dietary supplements in this country in 2007, of which $1.7 billion went for potentially hazardous weight-loss products.[5]

## Hatch-Harkin and the Regressive Era

Since the New Deal era, when the present FDA took up the regulatory torch of the Pure Food and Drug Act, the agency considered that dietary supplements were simply nutritional items such as vitamins, minerals and proteins. However, as New Age herbalists, ethnic botanists and ashram dwellers came of voting age, the Nutrition Labeling and Education Act of 1990 added "herbs, or similar nutritional substances" to the term dietary supplement.[16] The Internet and word of mouth could now peddle such agents of questionable composition. No tests of safety or toxicity were required. Bad news: a search in PubMed for "dietary supplements/toxicity" reveals 100 such reports since 1990, ranging from the heavy metal contamination of ayurvedic medicines to renal dysfunction following oral creatine supplements.[17,18]

Nevertheless, a huge para-pharmaceutical industry, which soon found friends in the legislature, was spawned. In the final hours of the 103rd Congress, the Dietary Supplement Health and Education Act (DSHEA) of 1994 passed easily, thanks to major bipartisan support. In the Senate, Orrin Hatch (R-UT) and Tom Harkin (D-IA) were its sponsors, and former Congressman Bill Richardson (D-NM) was its sponsor in the House. The legislators were assured that the bill was popular: "Congress has received more mail on this subject than any other issue, including Bosnia, the Gulf War, Somalia, gun control, tax reform, and healthcare reform!"[19]

The 1994 DSHEA, also called the Hatch-Harkin act, expanded the meaning of the term "dietary supplements" beyond essential nutrients to include such substances as "ginseng, garlic, fish oils, psyllium, enzymes, glandulars, and mixtures of these." Moreover, the DSHEA assured manufacturers that they could describe the supplement's effects on "structure or function of the body [or the] well-being achieved by consuming the dietary ingredient" without substantial scientific proof; nor were the supplements required to warn against possible side effects in the same manner of prescription drugs.[19]

You don't need to be a pharmacologist to suspect that almost anything that *really* affects the structure or function of the human body might have an unwanted side effect (aka, toxicity). Indeed, a search in PubMed for "herbal drugs/toxic effects" finds such 460 articles since Hatch-Harkin. These range from hepatotoxicity from herbals and weight-loss supplements in the United States[20] to kidney failure from use of aristolochia, a Chinese herb used worldwide.[21] Announcing passage of the 1994 bill, the legislators were proud that "the nutritional supplement industry is an integral part of the economy of the United States."[22] It was a statement that could have been found in *Tono-Bungay*:

> *Yes, I thought it over—thoroughly enough—Trade rules the world. Wealth rather than trade! The thing was true, and true too was my uncle's proposition that the quickest way to get wealth is to sell the cheapest thing possible in the dearest bottle.*[23]

Hatch-Harkin also provided that "unlike health claims, nutritional support statements need not be approved by the FDA before manufacturers make products bearing these statements."[22] That's why you see so many claims for supplements that promote "joint health," "breast health" or "male vigor" rather than more precise claims that would have to be validated by the FDA in following ethical pharmaceuticals. In fact, when Wallace et al.[24] applied FDA "adverse report standards" to dietary supplements, they found at least 472 adverse events in the period between 1999 and 2003 caused by echinacea, ginseng, garlic, ginkgo biloba, St. John's wort and peppermint. They worried that lax "U.S. policy on dietary supplements stands in contrast to the more rigorous approach of the European Union."[24] Worse yet, lack of regulation made it possible for manufacturers to spike their herbals with sometimes toxic prescription drugs, such as bumetadine, phenylbutazone and aminopyrine.[3,25]

It is clear to me that the deregulation of pills and potions containing mixtures of questionable ingredients is another aspect of the general notion that "government is the enemy," an idea that has been as productive in the subprime drug industry as in the subprime mortgage industry. Hatch-Harkin advocates spelled out their reasons for the absence of warning labels on dietary supplements: "Some manufacturers were fearful that putting responsible warning labels on some herbal products might create a situation in which FDA would regulate the product as a drug, since food products do not generally contain warnings."[19] That's almost a summary of the Regressive Era of the last decade and a half.

In *Tono-Bungay*, the snake-oil millionaire goes bust after having overextended his empire. His nephew, George, who survives financial disaster, goes on to fantastic adventures in science and technology. Looking back at the world of patent elixirs, he issues a dark warning against the perils of unregulated commerce:

*I did not realize until after the crash how recklessly my uncle had kept his promise of paying a dividend of over eight per cent on the ordinary shares of that hugely over-capitalized enterprise. . . . The flags flutter, the crowds cheer, the legislatures meet. Yet it seems to me indeed at times that all this present commercial civilisation is no more than my poor uncle's career writ large, a swelling, thinning bubble of assurances; that its arithmetic is just as unsound, its dividends as ill-advised, its ultimate aim as vague and forgotten.*[26]

# Voodoo Economics and Voodoo Healing: Witchcraft Persists in Massachusetts

*Pope Benedict XVI, nearing the end of his first pilgrimage to Africa . . . told priests and nuns of their duty to divert their fellow Angolans from malign beliefs in witchcraft and sorcery.*

—*The New York Times*, 2009[1]

*A trained practitioner of Reiki healing has the ability to channel pure life force—what is known as ki, chi, or prana—and gently allow the recipient to absorb this life force. . . . As a practicing witch, I use healing magick in my life all the time.*

—Christopher Penczak, 2005[2]

*Reiki is a simple, natural and safe method of spiritual healing and self-improvement that most everyone can benefit from. It has been found to be effective in helping every known illness, including cancer.*

—Dana-Farber Cancer Institute, 2009[3]

*It just isn't going to work, and [is] what I call a voodoo economic policy.*

—George H. W. Bush, 1980[4]

### Bewildered in Angola

WHEN POPE BENEDICT LEFT AFRICA IN 2009, he rebuked Angolans for cruelly persecuting their fellow citizens. "In their bewilderment they even end up condemning street children and the elderly as alleged sorcerers," the pontiff said.[5] Angola has the continent's highest percentage of Catholics, yet much of the population retains its ancient belief in the power of witches and sorcer-

70

ers. Angolan voodoo practices mix fetish worship with Catholic symbols, and this national "bewilderment" has been strengthened by brutal civil war and the AIDS epidemic. As a result, children and elderly folk accused of witchcraft have been burned, mauled and mutilated for casting spells that cause family misfortunes—from dysentery to divorce, malaria to myopia. Traditional healers, accredited by the government, join with Pentecostal clerics to exorcise spells cast by under- or over-age sorcerers. "One boy was kept in a chicken coop, and parasites ate out his eye. Another was found living in a toilet pit," the BBC reported.[6] A little girl, accused of casting evil spells, had her hands burned on a red-hot stove, her clothes were torched and she was beaten by her mother and sisters before a large crowd despite repeated pleas of innocence.[7] A senior Pentecostal pastor exorcised evil spirits from a starved and terrified, nearly naked girl of eight. Her head was shaven and foul matter had been rubbed into her eyes for several days. He dismissed any possibility that the child could die from mistreatment. The pastor explained: "Why should the child die? If the child dies, it means the child is evil."[6]

Evil or not, many children die in Angola and very few get to be elderly. The sad credulity of Angolans with respect to voodoo and witchcraft are matched by the country's dismal vital and economic performance. Today's life expectancy in Angola is 38.2, and although the average woman brings 6.2 children to term, at least one will die in infancy (180 infant deaths per 1,000 live births). The median age of Angolans is 18, and almost half of the population is under 15. The per capita income is $1,350.[8,9]

In the United States, neither children nor the elderly have been persecuted for witchcraft since that episode in Salem (1692). Our life expectancy is 78.2. The average American woman brings an average of two children to term and neither of these is likely to die in infancy (six infant deaths per 1,000). The median age of Americans is 37, and only 20 percent of the population is under 15. In 2009, the per-capita income in the United States was $39,751.[8,9]

From these data, we might conclude that nations "bewildered" by voodoo and witchcraft are unlikely to support prosperity or a long life span. But we'd be wrong, for witchcraft is making a comeback in well-off Massachusetts (per-capita income $50,735) and voodoo healing is alive and well at Harvard (perhaps the least bewildered university in the state).

### Witches of the World Unite

The Massachusetts homepage for latter-day witches, Witches of the World, welcomes witches, pagans, wiccans and heathens with the good news that the Massachusetts resource pages have been visited 1,289,582 times, as of April 15, 2009.[10] And the Harvard Medical School's *Health News Letter* recommends a form of touch therapy favored by witches, Reiki, as "among the more promising" approaches in cardiology:

> *Healing touch. Practitioners use their hands above or on the body, using a gentle touch, with the intent of affecting the body's energy field . . . Reiki. This centuries-old practice involves light touch over different parts of the body in an ordered sequence.*[11]

Reading this twaddle on a spring day in Woods Hole, Massachusetts, it seemed to me that Pope Benedict could well issue a warning against "bewilderment" in Massachusetts. The Bay State is positively awash in witches, sorcerers and Reiki fanciers.

At one gateway to Cape Cod, the Plymouth Area Witches (membership 37) urge their clan to make "magickal friends" to "see spells and witches come true!" Therapeutically inclined, they claim that while their main focus is on healing and divination, they will also share spells, meditations and "experiences."[12] At Rehoboth, Massachusetts, the other entry to the Cape, "The coven grew until we no longer had room for new members (though our sabbats were always open to non-coven members)."[13]

Slightly to the north, an event took place that Pope Benedict might have good reason to abjure. On April 10, in Spencer Massachusetts, the practicing witch, Christopher Penczak, presided over a meet-and-greet book signing for his *Magick of Reiki: Focused Energy for Healing, Ritual and Spiritual Development*. He explains in his writings that

> *Reiki also works on the energy bodies. Healing can occur on an emotional, mental, or spiritual level. Reiki differs from other forms of energy healing, because the recipient is said to be in charge of the transfer of energy. The recipient's "higher self" or "body wisdom" controls the flow, preventing any overload from too much energy. The Reiki practitioner is like a straw, the recipient sucks energy through the straw, and just like having a drink, when you've had enough, you stop drinking. The recipient cannot "choke" or overload on the healing energy.*[2]

Mr. Penczak differs on this last point with the authorities at Harvard's Dana-Farber Cancer Institute, who assure visitors to their website that although Reiki promotes healing by "stimulating the immune system," there are—so help me—"Possible Side Effects." Those listed include "emotional release (laughter, crying, etc.), gastric upset, mild nausea, diarrhea, body sensations like tingling feeling, headache, etc."[3]—not unlike those qualifications announced for psychotropic drugs in the midst of the evening news.

The Harvard medical faculty seems to have a soft spot for this ancient therapy favored by witches. Writing in the authoritative *The Journal of Alternative and Complementary Medicine*, Catherine E. Kerr and her colleagues have carefully analyzed—and praised—the animal magnetism of the Reiki practioner:

> *He/she then sweeps his/her hands at a distance of 1–2 inches from the body and tells the patient he/she is receiving a powerful energetic touch. Even though the practitioners' hands are kept at a distance, patients frequently describe a "flowing feeling" near where they envision the practitioner's hands to be.*[14]

The Harvard group is not quite as bold as the National Center for Complementary and Alternative Medicine (NCCAM), which attributes to Reiki a dazzling power to act at a distance: "Practitioners with appropriate training may perform Reiki from a distance, that is, on clients who are not physically present in the office or clinic."[15] If that ain't voodoo, I don't know what is.

Not to worry, though. Harvard University's Pluralism Project is on the job. When Mississippi victims of Hurricane Katrina needed physical and spiritual help, they turned to local witches and "pagan" rites. Grove Harris, managing director of the Pluralism Project at the time, reassured network news that magic is the "art of changing consciousness at will. . . . It's not the same thing as some of the TV shows that portray a witch winking her nose and creating physical change. The way pagans can work with energy and clinical magic, in some ways it's similar to prayer," Harris said.[16]

Like rabies, Reiki and witchcraft have crossed the Cape Cod Canal. There is an active Witch School in Provincetown "dedicated to supporting Magick Media in all it's [sic] forms."[17] And, one can obtain Magickal Beads in West Falmouth, a stone's throw—so to speak—from the busy labs at Woods Hole.[18] Where Magick thrives, there lives Reiki. In nearby Dennisport one can consult a "Kabbalistic Healer and Reiki Master" whose experience with "craniosacral, hot stone, reflexology, and aromatherapy" promises the "relief from stress and pain."[19] This sort of thing may be OK coming from a day spa, but it's troublesome to have it sound like medical advice from the Harvard Health Letter or from Dana-Farber's website.[3,11]

## The Daughter of Fancy or Terror

Witchcraft in Massachusetts today is a far tamer affair than in Angola or in old Salem: children and the elderly are not hunted, burned or persecuted for sorcery. Magickal beliefs in the Bay State can be attributed to personal whim or fancy, while credulity in Africa remains a creature of terror. The distinction between those two faces of credulity was first drawn by James Russell Lowell in an 1868 essay on the Salem witch trials of 1692:

> *Credulity, as a mental and moral phenomenon, manifests itself in widely different ways, according as it chances to be the daughter of fancy or terror. The one lies warm about the heart as Folk-lore, fills moonlit dells with dancing fairies, sets out a meal for the Brownie . . . and makes friends with unseen powers as Good Folks; the other is a bird of night, whose shadow sends a chill among the roots of the hair: it sucks with the vampire, gorges with the ghoul and commits uncleanliness with the embodied Principle of Evil, giving up the fair realm of innocent belief to a murky throng from the slums and stews of the debauched brain.*[20]

As in Angola today, witchcraft in Salem was a true daughter of terror. Clerics were up in arms; the very young and the very old were in constant jeopardy of life and limb for casting evil spells. Salem's Rev. Parris had held Satan and his coven responsible for strange fits suffered by his daughter and niece, and soon the town was aflame in mutual recrimination. In consequence, hundreds were accused, 19 of the alleged witches were hanged, and poor Giles Corey, age 81, was pressed to death beneath a weight of stones for failure to plead. On August 19, 1692, George Jacobs, age 71, was hanged, still pleading innocent: "Because I am falsely accused. I never did it." He was hanged for having pinched, scratched and bitten a neighbor, "threatening to destroy him if he did not sign the devil's book." George's principal accuser was his own granddaughter, who accused George in order to save her own life.[21]

Lowell pointed out that "although some of the accused had been terrified into confessing, yet not one persevered in it but all died protesting their innocence, though an acknowledgment of guilt would have saved the lives of all." In Salem the accused were also not, as commonly the case elsewhere, abandoned by their friends. "In all the trials of this kind," he wrote, "there is nothing so pathetic as the picture of Jonathan Cary holding up the weary arms of his wife during her trial, and wiping away the sweat from her brow and the tears from her face."[20]

Lowell tells how a poor slave girl, Tituba, and a score of other defendants in the witchcraft trials were forced to confess to casting spells and to having "out-of-body" experiences, to seeing great balls of fire, to having visions of multitudes in white glittering robes and the sensation of flying. Closer to fancy than terror, Tituba's confessions seem to ring a modern bell (and book and candle, perhaps?); they antedate by three centuries the "experiences" promised by the Witches of Plymouth or the Magick of Reiki.[2,12]

## Voodoo Economics

Of that proud antebellum company of social reformers and abolitionists who had made Boston the temporary Athens of America, Lowell was perhaps the most cosmopolitan. He succeeded Longfellow as professor of modern languages at Harvard; he was the first editor of *The Atlantic Monthly* and *The North American Review*. He later became American minister to Spain—where he discovered Goya's dark images of war—and to the Court of St. James. The poetry and criticism he wrote were good enough for contemporaries to rank him among the lions of the Saturday Club (that is, Emerson, Longfellow, Whittier, Holmes and Hawthorne).[22]

But Lowell was also an astute social critic. He noted that many of those hanged or pressed to death were elderly, and many of the accusers were young children. In the Salem of 1692—as in Angola today—it was rare for people to reach the age of 70 or 80. Lowell described that Old Folks were regarded as curiosities, biological outliers who must have made a pact with the devil: "Unhappily there were always ugly old women; and if you crossed them in any way, or did them a wrong, . . . they could send a demon into your body, who would cause you to vomit pins, hair, pebbles, or knives."[20] In the land where witchcraft reigned, life was short. Life expectancy in Puritan Massachusetts was only a little better—slightly over 40—than for Angolans today. In the Bay State today, in our age of stents, beta-blockers and the ICU at the MGH, we've doubled that short span of years.[23] We owe that to reason, not placebos.

Lowell also traced the Salem witchcraft trials to socioeconomic causes. He attributed Salem's obsession with witches to its roots in Cromwellian doctrine and argued that earlier witchcraft trials in England were an aspect of Calvinist belief. The trials were grounded in a tradition of contractual agreements among the mercantile class. Lowell suggested that Christianity had invented the soul as an individual entity to be "saved" or "lost"—like money—and that the grosser wits of commerce came to treat the soul as a "piece of property that could be transferred by deed of gift or sale, duly signed, sealed and witnessed." He argued that the Salem model of witchcraft was based on an unholy contract between an individual witch and the Evil One.[20] The success of their joint venture was to be judged by the number of souls they had pocketed: the ultimate bottom line, we might say.

The trade that was in the saddle in Salem was traffic in souls. The trade in witchcraft and Reiki today also has economic consequences. The NCCAM has spent over half a billion dollars in the last four years to fund voodoo sciences such as Reiki, aromatherapy and homeopathy—practices that the NIH itself admits may not "follow the laws of science (particularly chemistry and physics)."[24] That's money taken from the Harvard Medical School or the Dana-Farber Cancer Institute where cures have been discovered by methods that do in fact follow the laws of science, particularly chemistry and physics. It is also reassuring to learn that a recent, serious clinical trial of Reiki for fibromyalgia showed that the laws of chemistry and physics hold up: Reiki is entirely useless.[25] We might conclude that it's also useless to put any further money into voodoo healing that works via the "life-force" of animal magnetism. In the prophetic words of George H. W. Bush on Reaganomics: "It just isn't going to work, and [is] what I call a voodoo economic policy."[4]

# Myrna Loy: Co-Principal Investigator

*"A Bacardi," says Nick to the waiter. He looks to Nora, who nods. Nick motions to the waiter and adds, "Two Bacardis." Says Nora with a straight face to the waiter, "I'll have the same." The waiter returns with four Bacardis.*

—*Another Thin Man*, 1939[1]

*Great women scientists are invisible until someone specifically looks for great women scientists. Then they are plentiful. But no one has to look for great male scientists. Great male scientists always appear on lists of great scientists.*

—Jo Handelsman and Rosalind A. Grymes, 2008[2]

*Why does every black person in the movies have to play a servant? How about a black person walking up the steps of a courthouse carrying a briefcase?*

—Myrna Loy to MGM Executives, ca. 1934[3]

*I cannot and will not cut my conscience to fit this year's fashions.*

—Lillian Hellman in 1952, *Scoundrel Time*, 1976[4]

## Looking for a Few Good Women

IT'S BEEN A LONG AND WEARISOME SEASON in American politics, a scoundrel time in which issues of race and gender have been crudely exploited. Yet, whatever the outcome was of the 2008 elections, two barriers to democracy in America have surely been breached. Never before has a black man carried a briefcase so high up the courthouse steps, never before has the glass ceiling been so effectively challenged. Cynics may gloat—"Guns, God, Lipstick?

This is what feminism has come to?"[5] or "If Obama was a white man, he would not be in this position"[6]—but the rules of the political game have changed—forever, one would hope.

At least with respect to gender, the rules have also changed in science and I'm going to argue that the silver screen is at least partly responsible. The notion came to mind early this fall, when a quick run-through *The Thin Man* series on Netflix coincided with a sparkling editorial in the journal *DNA and Cell Biology*. Titled "Looking for a Few Good Women?," it asked why women have been short-changed in the glittering prizes of science. Its authors, Jo Handelsman of the University of Wisconsin and Rosalind Grymes of the University of California Santa Cruz, point out that women have received almost half of all U.S. doctoral degrees awarded in the life sciences over the last 20 years. They constitute 32 percent of all tenure-track faculty and 26 percent of full professors in the life sciences.[2] That's one part of the story.

Handelsman and Grymes, officers in the vigorous Rosalind Franklin Society,[7] correctly point out that women are underrepresented in our academies and among winners of the Lasker and Nobel Prizes. Their data, with some additions, are presented in Table 1. The steeper the slope, the higher the prize, the tougher it is for talented women, who now form a large pool, to gain the glittering prizes. "What can be done?" they ask, and suggest that "We need to look for great women scientists, and if we use a rigorous process of review, we will see them. Lots of them."[2]

I was suddenly concerned that *The FASEB Journal*, a journal for which I am responsible, might not be keeping up its part in gender neutrality: we should have pretty much as many women as men among our authors.

**Table 1:** *Gender Distribution of Honor Awards in the Life Sciences*

|  | Male | Female | % Female |
|---|---|---|---|
| Nobel Prizes (Phys. or Med.) 2000–2007 | 20 | 1 | 5 |
| Lasker Foundation (Basic Sci.) 2000–2008 | 35 | 2 | 5 |
| National Academy of Sciences 2005–2008 | 232 | 56 | 19 |
| Ellison Senior Fellows 2000–2008 | 112 | 25 | 18 |
| Howard Hughes Investigators 2005–2008 | 74 | 25 | 25 |
| Ellison Junior Fellow 2000–2008 | 92 | 36 | 28 |
| NIH Director's Pioneer Award 2004–2007 | 93 | 34 | 28 |

Data from References 2 and 8.

**Table 2:** *Authorship and Editorship in Life Science Journals*

|  | Male | Female | % Female |
|---|---|---|---|
| *FASEB J.* authors (Aug.–Sept. 2008) | 188 | 166 | 47 |
| *FASEB J.* editors (multiple) | 43 | 14 | 25 |
| *J. Exper. Med.* authors (Jan.–Dec.1933) | 125 | 22 | 15 |
| *J. Exper. Med.* editors (S. Flexner, P. Rous) | 2 | 0 | 0 |

And since our editors are all full professors, at least a quarter of our editors should be female.

In Table 2 are listed the gender of all authors of refereed articles in the August and September issues of the journal, in so far as our imperfect knowledge of gender-specific first names permits. I compare this to a comparable journal, *The Journal of Experimental Medicine* of 1933, the year Dashiell Hammett's *The Thin Man* was written—the film appeared a year later. The *JEM* is a reasonable control, since at the time it published the sort of broadly interdisciplinary articles that *The FASEB Journal* has been publishing only since 1987.

Today's numbers may not be entirely satisfactory, but in the life sciences, women are better represented than in the days of speakeasies and breadlines and of journals controlled by one or two very eminent seniors. And yet . . .

It was clear to Handelsman and Grymes, and painfully evident to me as well, that from this ample supply of good to great women scientists (25 percent of Hughes Investigators!), perhaps more than one or two female laureates might have been honored at dinner in Stockholm or lunch in Manhattan. One should have thought that a pool of fine Investigators is waiting to be tapped.

### I'll Have the Same

That's where Myrna Loy comes in. She broke the glass ceiling in the private investigator game, and went on to play an honorable role in the scoundrel time of the McCarthy Era. Myrna Loy, and the couple who inspired the characters of Nick and Nora Charles in *The Thin Man* series—Lillian Hellman and Dashiell Hammett—can remind us of the contribution made to gender equality by the progressives in our midst. They were part of the Popular Front in the New Deal era, crusaders for social justice and family planning, for contraception and reproductive research, for women's education and childcare. Without those "pinkos" and "fellow travelers," as Joseph McCarthy called them, there'd have been far fewer opportunities for women to leave their *Kinder, Küche und Kirche* (kids, kitchen and church) to pursue careers in the life sciences.

Hellman and Hammett were clearly the models for Nick and Nora, and they spoke to each other as such: Hammett wired from Hollywood to Hellman in New York as he was adjusting *Another Thin Man*: "SO FAR SO GOOD ONLY AM MISSING OF YOU PLENTY LOVE NICKY."[9] They surely would be welcomed by today's Rosalind Franklin Society, happy that Nick and Nora Charles "recognized and fostered" the image of female competence on the screen.[10] The poet Frank O'Hara had it right: it wasn't *The Nation, The New Republic* or *Partisan Review* that carried the torch for the Popular Front and equal recognition, but the movies:

> *Not you, lean quarterlies and swarthy periodicals*
> *with your studious incursions toward the pomposity of ants,*
> *. . . but you, Motion Picture Industry,*
> *it's you I love!*
> . . . Myrna Loy being calm and wise, William Powell
> in his stunning urbanity. . . .[11]

The Myrna Loy who in six films of *The Thin Man* series played Nora Charles was probably a more effective messenger of progress than the Myrna Loy who fought to abolish the House Un-American Activities Committee. Loy and Powell cast the mold for other male–female pairs of equal wit and wisdom: Hepburn–Tracy, Rogers–Astaire and Bacall–Bogart. At a Lincoln Center tribute to Myrna Loy, Lauren Bacall acknowledged that debt and asked:

> *And to meet whom did Franklin D. Roosevelt find himself*
> *tempted to call off the Yalta Conference? Myrna Loy. And to*
> *see what lady in what picture did John Dillinger risk coming*
> *out of hiding to meet his bullet-ridden death in an alley in*
> *Chicago? Myrna Loy, in* Manhattan Melodrama.[12]

Her wisdom was not limited to screenplays. FDR named her an assistant to the director of military and naval welfare for the Red Cross in World War II. She was later appointed a member-at-large of the U.S. Commission to UNESCO.[13]

### Good Enough for a Detective, but Not for Me

Before Nick and Nora Charles emerged as co-investigators, detective fiction and "mystery" stories were dominated by omniscient males. The great detectives used wit and whimsey as weapons; they were never simply funny. Arthur Conan Doyle's Sherlock Holmes (1887), Agatha Christie's Hercule Poirot (1920), Dorothy Sayer's Lord Peter Wimsey (1923) and H. C. McNeile's Bulldog Drummond (1920) ruled the genre with guile and style; Miss Marple appeared as a village snoop in 1927. It wasn't until Dashiell Hammett himself broke the mold with Sam Spade (1930) that irony became de rigueur. Nick and Nora, on page and on screen floated, flirted, drank and joked:

> NICK:   *"I'm a hero. I was shot two times in the Tribune."*
> NORA:   *"I read where you were shot five times in the tab-loids."*
> NICK:   *"It's not true. He didn't come anywhere near my tabloids."*[14]

And when it came to investigation, Nora set the mold for how an investigator differs from a detective. In the original *The Thin Man* book, Hammett has Nora set out the method as Nick nods off:

> *Nora yawned again. "That may be good enough for a detective, but it's not convincing enough for me. Listen, why don't we make a list of all the suspects and all the motives and clues, and check them off against—" "You do it. I'm going to bed."*[15]

It may be no accident that Nick became a private "investigator"—Hammett had worked as a Pinkerton detective before he took up his pen. Like Hammett, Nick Charles has retired from being a detective for hire, and now he and Nora pursue crime as independent *investigators* rather than working for hire as *detectives*, a police rank. (Those of us in academic science can appreciate the difference: you have to raise your own money as an investigator.) It was said of Hammett, and of Nick Charles, that he was doing "an intensive study of the liquor problem from the consumer's standpoint."[16] Nick and Nora were the models of co-principal investigators with more than a touch of class. They were also models of urban charm and sophistication, mixing martinis with wisecracks on the silver screens of the world.

In contrast to the neat gloss of Nick and Nora, Hellman and Hammett had messy personal lives. Both were boozers, chain-smokers and contentious. Their literary careers were troubled by rivalry, enmity and betrayals. John Leonard summed up their messy affairs:

> *Never mind their talent. We are talking about Hollywood. It's now my view that one promiscuous drunk found another to*

*reciprocate, that they spent thirty years jockeying for the perfect*
*vampire-bite position, and that, as their lips locked, so did*
*their wounds.*[17]

But they both had moments of great courage. Hellman received a sub-poena to appear before the House Un-American Activities Committee in 1952. She was asked to disclose the names of her associates in film and the theater who could have been Communists or fellow travelers. She refused to comply and published her response in *The Nation*:

> *To hurt innocent people whom I knew many years ago in order*
> *to save myself is, to me, inhuman and indecent and dishonor-*
> *able. I cannot and will not cut my conscience to fit this year's*
> *fashions, even though I long ago came to the conclusion that I*
> *was not a political person and could have no comfortable place*
> *in any political group.*[4]

Well, not really: she was indeed a convinced fellow traveler. So be it. But in consequence, Hellman's name wound up on Hollywood's blacklist and she was stuck with an inexplicable bill by the IRS. She survived and her plays such as *The Little Foxes* and *The Children's Hour* continue in perfor-mance the world over.

A disabled veteran of World War I, Dashiell Hammett again enlisted in the Second World War at the age of 48, became a sergeant and was stashed away in the Aleutians, suspected of "pre-mature anti-Fascism." After the war, he was a member of the Civil Rights Congress, a liberal political group that was targeted by the FBI as being a communist front. When he refused to name contributors to the organization and was sentenced to six months in jail. In a 1953 episode that has resonances with the political wars of 2008, Sen. Joseph McCarthy asked him if the government should fund purchases for library books written by avowed communist sympathizers. Hammett replied, "If I were fighting Communism, I don't think I would do it by not giving people any books at all."[18] His literary career dissolved in chronic lung disease and booze, and he died penniless in Katonah, New York, in 1961. According to his wishes, this "pinko" father of Nick and Nora, the lover of Lillian Hellman, was buried in Section 12 of Arlington National Cemetery. FBI Director J. Edgar Hoover attempted to block the burial but was overruled by the Dept. of Veterans Affairs.[18] His simple gravestone reads: SAMUEL D HAMMETT MARYLAND TEC 3 HO CO ALAS-KAN DEPT WORLD WAR I & II MAY 17 1894 JANUARY 10 1961.

# Dr. Ehrlich and Dr. Atomic:
# Beauty vs. Horror in Science

*Sometimes it seems as if horror is the only story that science has to tell, or the only one we want to hear. Somebody has a gadget they have to build, an experiment too sweet to resist. . . . The tug of war between beauty and horror is the theme of* Doctor Atomic.

—Dennis Overbye, *The New York Times*, 2008[1]

*When I was a boy, I looked out into the star-filled sky one night and was awestruck by its beauty. . . . A few days later, I happened on a book called* The Microbe Hunters *and became equally enchanted by the stories of microbes and their role in disease. It dawned on me that I wanted to explore this hidden universe.*

—Stanley Falkow on his Lasker Award
in Medical Research, 2008[2]

## I Kid You Not. . .

IT's A RARE EVENT WHEN a neglected work of popular literature, Paul de Kruif's *The Microbe Hunters*, is linked to the birth of recombinant DNA. It's also a rare event when a 17th-century sonnet, John Donne's "Batter My Heart," becomes an aria in a new opera sung by a poetically inclined physicist at the birth of the atom bomb. Both events took place in mid-Manhattan in the fall of 2008, and the coincidence is more than geographic. It was also mid-election. One recalls that Paul Ehrlich, the Microbe Hunter, and J. Robert Oppenheimer, of "Batter My Heart," underwent shameful public trials fueled by notions resurrected in 2008 by Joe the Plumber. Ehrlich and Oppie became targets of nativist, neo-Luddite rhetoric directed not only against their persons but against science itself: "I kid you not" as they say in the sub-Arctic.[3]

## Microbe Hunters at the Lasker Awards

Stanley Cohen evoked Paul de Kruif's 1926 book *The Microbe Hunters* as he presented a 2008 Lasker Special Achievement Award in Medical Research to his Stanford colleague, Stanley Falkow. He hailed Falkow's discovery that bacterial plasmids determine antibiotic resistance and virulence, explaining how they had made the revolution in molecular biology possible. Cohen proposed his colleague for a new "pantheon of great Microbe Hunters," recalling that in 1972 Falkow contributed to a discussion at a Waikiki beach delicatessen "over corned beef and pastrami sandwiches and very cold beer" that resulted in the invention of recombinant DNA by Herb Boyer and Cohen himself.[4]

Evoking that now legendary birth of biotechnology in Hawaii, Falkow traced the roots of his own career to the heroes of de Kruif's *The Microbe Hunters*—Pasteur, Koch, Dr. Ehrlich and his Magic Bullet et al.—who "became my heroes, and I dreamed of becoming a bacteriologist, doing research on the bacteria that cause disease."[1] Falkow's discovery of how R plasmids help bacteria dodge today's magic bullets made his dreams come true. He also played a role in the 1974 Asilomar conference[5] at which the first whistle of bioethics was blown on gene splicing, and served on the first NIH Recombinant DNA Advisory Commission.[1] His place seems secure in that new "pantheon of great Microbe Hunters."

Paul Ehrlich is well established in the older pantheon: he made three discoveries that opened up new fields in science. He was the first to classify cells and microbes according to their affinity for azo-dyes (cytochemistry), he gave weight and number to toxins and antitoxins (immunochemistry) and he developed Salvarsan, that "magic bullet," as a cure for syphilis (chemotherapy).[6] Ehrlich won his Nobel in 1908 for work on humoral antibodies, the mediators of acquired immunity. He shared the prize with Ilya Metchnikoff, who discovered that phagocytosis was the basis of cellular, or innate, immunity.[7] Ehrlich elaborated his prescient model of how toxins interlock with their antitoxins, the side-chain theory of humoral immunity. He concluded that these lock-and-key reactions obey the laws of chemistry and physics, a reductive notion that did not endear him to the nativist philosophers of *Geist* (spirit). Against the Geist-hunters, he proposed his own "four Gs" as the path for scientific achievement: "*Geld, Geduld, Geschick und Glück*" (money, patience, skill and luck).[6]

The discoveries of three other 2008 Lasker awardees—Victor Ambros, David Baulcombe and Gary Ruvkun—are prime examples of why Ehrlich's

4 Gs beat the babble of *Geist* every time. The "unanticipated world of tiny RNAs that regulate gene function in plants and animals" is a world of lock-and-key reactions. You can't get more reductive than those micro RNAs, the magic bullets of molecular biology.[8]

## The Microbe Hunter on Trial

Two years after the notoriety of his Nobel Prize, and within a year of Salvarsan's use in the clinic, attacks on Paul Ehrlich ended in calumny and the courts.[9] The German vigilante pack was led by a nationalist physician, Dr. Richard Dreuw of Berlin, and a zealous Frankfurt pamphleteer, Karl Wassman, a "strange-looking man dressed in a dark monk's habit, with a rope around his waist . . . who believed in curing all diseases by Nature alone."[6] Wassman's pamphlet *Die Warheit* (The Truth) accused Ehrlich and his Japanese coworker Sahatschiro Hata (1873–1938) of concocting a dangerous, unreliable drug (called 606) and the Frankfurt hospital of shoddy record keeping. *Schadenfreude* and racism became an integral part of the story: "*Die fachliche Kritik an dem Heilmittel wird mit antisemitischen Angriffen auf Ehrlichs Person verbunden.*" (Technical critique of the drug went hand-in-hand with anti-Semitic attacks on Ehrlich himself.) Lutheran clerics argued that the wages of sin was syphilis and that Ehrlich et al. were disrupting the natural order of crime and punishment.[10]

Other familiar notes were sounded by populists. Ehrlich had signed over manufacture of the drug to the Hoechst Corporation, which charged ten Marks—60 dollars—for a course of Salvarsan. The critics complained that Ehrlich was getting rich, that Hoechst was profiting from basic research funded by the government and that clinical trials of Salvarsan had been carried out on the prostitutes of Frankfurt without their consent. Things came to a head in a long, drawn-out libel suit brought by the hospital against Wassman on behalf of Ehrlich and Hata. The proceedings turned into a circus as inflammatory witnesses for the defense were corralled from the red-light district and shadier areas of town. In the end, however, Wassman lost and was sent to prison and a worn-out Ehrlich was finally exonerated. But soon the First World War supervened and the Guns of August 1914 silenced the uproar in Frankfurt. Ehrlich died a year later, spirits shattered by the public scandal. Wassman was pardoned, changed the name of his pamphlet to *Die Liebe* (Love) and never mentioned 606 again.[6] Salvarsan went on to set the gold standard for the treatment of syphilis until 1937.

Sentiments that had fueled the Ehrlich trial re-emerged in the Nazi era. In August 1938, the Nazis removed the street sign for Paul Ehrlich Strasse in Frankfurt. The 1940 Warner Brothers' film *Dr. Ehrlich's Magic Bullet* was Hollywood's response to the year of Kristallnacht. The film's screenwriter, Norman Burnside, admitted that "the reason for picking Ehrlich as a protagonist had very little to do with syphilis and its cure," and its producer Hal Wallis agreed that the film was a visceral response to Hitler's 1938 diatribe in which he proclaimed that "a scientific discovery by a Jew is worthless." The 2008 Lasker Awards in the year of Obama were a measure of what defeat of the Nazis made possible. They were also a measure of why the atom bomb was built.

## Atom Bomb at the Met

Three weeks after the Lasker Awards ceremonies at the Hotel Pierre, John Adams's opera *Dr. Atomic* made its debut at the Metropolitan Opera, just across Central Park from the hotel. The setting of the opera is Los Alamos in July of 1945 as the first atom bomb is about to be tested. Adams casts J. Robert Oppenheimer, the lab director, as a Faustian hero "equally in love with the Bomb and his own inscrutability."[2] The music is stunning, the sets are striking and the libretto by Peter Sellars blends Oppenheimer's favorite poetry—John Donne, Muriel Rukeyser, Charles Baudelaire—with actual wartime texts. Beauty mixes with horror in two astonishing acts. Adams celebrates completion of the "Fat Man," an implosion-design plutonium bomb, in what critic Alex Ross describes as an "inexplicably lovely choral ode to the bomb's thirty-two-point explosive shell, with unison female voices floating above lush string and wind chords and glittering chorus of chimes and celesta."[11] The climax of the drama is the explosion of the Fat Man over a desert site named "Trinity" by Oppenheimer in homage to the three-personed God of John Donne's sonnet "Batter My Heart." In *Dr. Atomic*, the poem is set as an aria for the Age of Anxiety:

> *BATTER my heart, three person'd God; for, you*
> *As yet but knocke, breathe, shine, and seeke to mend;*
> *That I may rise, and stand, o'erthrow mee,'and bend*
> *Your force, to breake, blowe, burn and make me new.*
> *I, like an usurpt towne, to'another due,*
> *Labour to'admit you, but Oh, to no end,*
> *Reason your viceroy in mee, mee should defend,*
> *But is captiv'd, and proves weake or untrue. . . .*[12]

As the bomb explodes and the lights go out, Oppenheimer evokes Vishnu in the *Bhagavad-Gita*: "Now I am become Death, the destroyer of worlds." *Dr. Atomic* closes with the amplified cries of Japanese bomb victims echoing over a pitch-dark auditorium. The tug of war between beauty and horror ends in horror.

Oppenheimer had assumed the task of building the bomb as head of the Manhattan Project's Weapons Design and Research laboratory in Los Alamos. The lab was in a race to build an atom bomb before the Nazis had gotten theirs, and Oppie had recruited a group of the most accomplished physicists of his day, including Hans Bethe, Richard Feynman, Enrico Fermi, Edward Teller and Victor Weisskopf, along with scores of others. As things turned out, by 1945 the Nazi bomb project had fizzled, while in Los Alamos success became imminent.[13] But the bomb had grown into an attractive technical problem of its own. Oppenheimer recalled later how one could become blinded to horror:

> *It is in my judgment in these things that when you see some-*
> *thing that is technically sweet you go ahead and do it and*
> *you argue about what to do about it only after you have had*
> *your technical success. That is the way it was with the atomic*
> *bomb.*[14]

### Dr. Atomic on Trial

It's difficult, in retrospect, to know whether it was the technical "sweetness" of the project that persuaded Oppenheimer to agree to drop an atomic bomb on civilian populations. But, he persuaded himself that a dropped bomb was by no means certain to explode and that an unexploded bomb could perhaps be turned against America by an enemy. He worried that an advance warning might prompt the enemy to move POWs to the area (as in Saddam Hussein's use of unwilling hostages in Gulf War I). Finally, he reasoned that no "demonstration" site would be as effective in forcing an end to war as those that Germany and Japan had used to show what "shock and awe" could accomplish: Guernica, Rotterdam and Pearl Harbor.

*Doctor Atomic* ends on the 16th of July, 1945, with the explosion at Trinity. On July 18, Japan refused an Allied ultimatum issued at Potsdam for unconditional surrender—no warning of a possible bomb accompanied the ultimatum. On August 6, a uranium bomb with a force of 15,000 metric tons of TNT destroyed Hiroshima and, no response being received, on August 9 the "Fat Man" was dropped on Nagasaki. The two bombs killed close to 250,000 people, almost all civilians. The war ended on August 10, 1945.[14,15]

The next stage in Oppenheimer's career was perhaps the most difficult. President Truman had awarded him the Presidential Medal of Merit in 1946 for his work at Los Alamos; a year later he was appointed director of the Institute for Advanced Study at Princeton, where he served until 1966. Simultaneously, he served as chairman of the General Advisory Committee of the Atomic Energy Commission (AEC) from 1947 to 1952. But soon pressure mounted to build the ultimate weapon, an H-bomb, a move supported not only by the military but also by scientists such as Edward Teller and a coven of "nuclear strategists."[16] Oppenheimer opposed it, as he opposed stockpiling more A-bombs: "I do not think that a country like ours can be based on the fear of what its people can do. . . . I have a sense of impending disaster and a sense of frustration."[17] But by 1953, the H-bomb was a reality, and Oppenheimer's resistance was considered unpatriotic. Major General Kenneth D. Nicholas of the AEC branded Oppenheimer as "the leader of a calculated movement in opposition to the hydrogen bomb program even after President Truman had decided as a matter of high national priority to go ahead with it."[18] This charge, joined to earlier charges that he had hobnobbed with communists and supported the "socialist" side during the Spanish Civil War, resulted in his being hauled before an AEC personnel hearing. The charges of the proceedings became headline news nationwide; the McCarthy-era press had no scruples about reminding folks that Oppenheimer's "communists and fellow travelers" included names like Isaac Folkoff, Max Friedman, Giovanni Lomanitz, Frank and Jackie Oppenheimer, William Schneiderman and Joseph Weinberg.[19,20] Contemporary press and radio accounts carried nativist overtones that echoed the Ehrlich trial (and the campaign rhetoric in the fall of 2008). On June 29, 1954, Oppenheimer's security clearance was revoked and his contract with the Atomic Energy Commission canceled:

> *The Atomic Energy Commission announced today that it had reached a decision in the matter of Dr. J. Robert Oppenheimer. The Commission by a vote of 4 to 1 decided that Dr. Oppenheimer should be denied access to restricted data. . . . Certain members of the Commission issued additional statements in support of their conclusions. These opinions and statements are attached.*[21]

A decade later the political tides had shifted. Anti–Vietnam war activists, honoring Oppenheimer's resistance to the hydrogen bomb and to nuclear stockpiling, made him a symbol of academic protest against the military. His example was frequently cited as biological scientists worried about their own Manhattan Project: recombinant DNA. Nobelist George Wald was one of many in that period who supported Oppie even as he told students striking against the war at MIT that "Dropping those atomic bombs on Hiroshima and Nagasaki was a war crime. . . . Our business is with life, not death."[22] The moral ambivalence of Oppenheimer and his Los Alamos colleagues toward the products of physical science were a goad to the organizers of the Asilomar conference itself. If the biologist's business is life, the question becomes how far life science dare go.[5]

One of Oppenheimer's heroes was Galileo Galilei, another physical scientist torn by the ambivalence of power. A passage in Bertolt Brecht's *The Life of Galileo*—written in August of 1945 after Hiroshima—sums up what could have happened, had not biologists developed something like a Hippocratic oath for science at Asilomar. Galileo regrets his abjuration, and the earth still moves:

> *As a scientist I had a unique opportunity. In my day astronomy emerged into the market place. Given this unique situation, if one man had put up a fight it might have had tremendous repercussions. Had I stood firm the scientists could have developed something like the doctors' Hippocratic oath, a vow to use their knowledge exclusively for mankind's benefit.*[23]

# Free Radicals Can Kill You:
# Lavoisier and the Oxygen Revolution

Mme LAVOISIER: *Imagine what it means to understand what gives a leaf its color! What makes a flame burn. Imagine!*

—C. Djerassi and R. Hoffmann, 2001[1]

*Oxygen is nothing other than the most salubrious and purest portion of the air, such that . . . it appears in an eminently respirable state more capable than the air of the atmosphere of sustaining ignition and combustion.*

—Antoine-Laurent Lavoisier, 1775[2]

*. . . though pure dephlogisticated air* [oxygen] *might be useful as a medicine, it might not be so proper for us in the usually healthy state of the body; for as a candle burns out much faster in dephlogisticated than in common air, so we might, as may be said, live out too fast, and the animal powers be too soon exhausted in this pure kind of air. A moralist, at least, may say, that the air which nature has provided for us is as good as we deserve.*

—Joseph Priestley, 1775[3]

*. . . there are good reasons for assuming that the changes produced by irradiation and those which arise spontaneously in the living cell have a common source—the OH and HO$_2$ radicals.*

—Denham Harman, 1956[4]

## An Oxidative Burst

I'M GETTING WORRIED ABOUT OXYGEN: not too little, but too much. We've known since Lavoisier that flames burn, metals rust and we take breath—all thanks to oxygen.[2] But we hadn't learned how oxygen excess does harm until 1954, when Rebecca Gerschman et al.[5] worked out that oxygen poisoning and x-irradiation share the property of producing oxygen-derived free radicals. Add one electron to oxygen, as by irradiation, and superoxide anion ($O_2^{-\bullet}$) is formed; that's a free radical anion. Add two electrons, and you get a bleach, hydrogen peroxide ($H_2O_2$), which is not a free radical but can readily react with $O_2^{-\bullet}$ to form the very reactive hydroxyl radical ($OH^\bullet$). Add any of these to living tissues, and they do damage.[5] Based on this chemistry, Denham Harman formulated his heuristic "Free Radical Theory of Aging."

> . . . In regard to aging, I felt that there had to be a basic cause which killed everything. Further, this basic cause should be subject to genetic and environmental influences.[6]

It is no accident that Gerschman's and Harman's laboratories were funded by the Atomic Energy Commission. As Harman later observed, the work "was of particular interest in 1954 because of concern over possible nuclear war."[6] Well, thanks to the Cold War and the last 50-odd years of research in the area, we've learned that Priestley was right about that burning candle.[3] We've learned that we do "live out too fast," and our animal powers are indeed "too soon exhausted" when we are bombarded by oxygen-derived free radicals, whether generated from without or within our own bodies.

The flood of publications dealing with "oxidative stress" has risen almost dramatically in the past few years.[7] Scanning papers with oxidative stress in the title, I checked out the online Web of Science. Sure enough, the oxidative burst had increased by a log order between 2005 and 2009 (Table 1). Over 15,000 articles about oxidative stress appeared in the literature in 2009, a 16-fold increase in only five years! In 2005, papers dealing with low oxygen (hypoxia) exceeded in popularity the hyperoxic condi-

**Table 1:** *Publications Per Year on Web of Science*

| Year | Oxidative stress | Hypoxia | Actin | RNA |
|------|------------------|---------|-------|------|
| 2005 | 906 | 4,277 | 4,938 | 23,497 |
| 2009 | 15,011 | 5,086 | 4,848 | 25,897 |

tion, but by 2009, articles about oxidative stress beat out work regarding hypoxia, by 3:1. Publications on ever-popular actin and RNA showed no significant change.[8]

Many of those papers spell out the fact that we age by the same mechanism by which metals rust, photos fade or wicker frays. They also confirm Harman's prediction that in our bodies, these processes are "subject to genetic and environmental influences."[6] We know that genetic defects in one or another of our own defenses against those free radicals, such as superoxide dismutase[9,10] or ceruloplasmin,[11] cause serious disease. Our cells also make reactive oxygen species, not only in the course of mitochondrial respiration within cells, as already surmised by Gerschman et al.[5] Our inflammatory cells also assemble machinery for making $O_2^{-\bullet}$ and $OH^{\bullet}$ on the cell surface.[12] In concert with a newly recognized brew of reactants formed from NO, ozone and so on, these wreak havoc with our cells and extracellular constituents.[6] Lavoisier had the chemistry right when he told us what happened to the animal oils of our cells when oxygen reacts:

> *It is evident that the oils, being composed of hydrogen and charcoal combined, are true carbono-hydrous or hydrocarbonous radicals; and, indeed by adding oxygen, they are convertible to vegetable oxides and acids according to their degrees of oxidation.*[13]

And that was before hydroperoxy and peroxy fatty acids—our century's version of vegetable oxides—were on our screens.

## A Revolutionary Painting

Antoine-Laurent Lavoisier and his wife Marie-Anne Lavoisier are depicted in the most beautiful image of scientific coworkers ever put on canvas. Jacques-Louis David's 1788 portrait of this unusual couple is not only an image of Enlightenment grace and grandeur but also a document of a work-in-progress for which both of the principals are responsible— the revolutionary *Traité Élmentaire de Chimie* of 1789.[13] It justifies the later appellation of Lavoisier as the "Father" and Mme. Lavoiser as the "Mother" of modern chemistry.[14]

Lavoisier is shown with quill on paper looking up at his wife, as if to take dictation or suggestion. The instruments on the table and floor are those used by Lavoisier to give "the first accurate accounts of burning, respiration and rusting."[15] Mme. Lavoisier is depicted with her arm on husband's shoulder while the folio behind her holds drawings for some of the 13 plates she fashioned for the *Traité de Chimie*. Lavoisier's mid-script attention to his wife alludes to her important translation of Richard Kirwan's polemical treatise on phlogiston, from English to French. In turn, her modest smile directed at the viewer—and the portraitist—suggests a strong student/teacher bond. Marie-Anne was an apt pupil of David: her skillful student drawings, with David's comments, remain extant in the Musée des Arts et Métiers in Paris.[16]

The portrait also documents a concurrence of three revolutions: the American, the French and the Chemical. It is set in the Arsenal of Paris, to which Lavoisier had been appointed as commissioner of the Royal Gunpowder and Saltpeter Administration. His charge, so to speak, was to greatly improve the purity and efficacy of French explosives. He fulfilled this task admirably: in aid of his American friends, Thomas Jefferson and Benjamin Franklin, he guaranteed the colonists a trusty supply of neat gunpowder to fight the Redcoats.[17] Indeed, Franklin was a good friend of both Lavoisiers. Recovering from an attack of gout back in Philadelphia, Franklin wrote to Marie-Anne in 1783 to thank her for a portrait she had painted of him for its "great merit as a picture in every respect; but what particularly endears it to me, is the hand who drew it."[18] He was only one of Marie-Anne's many admirers, who included Gouverneur Morris, P. S. du Pont de Nemours and Benjamin Thompson, Count Rumford. As Roald Hoffmann lamented: "There is no biography of Mme Lavosier. I think she deserves an opera."[15]

## A Chemical Revolution

Dispute has raged over which of the three chemists who came across oxygen between 1772 and 1778 deserves credit for the air of life. It is now agreed that a Swede discovered it first, the "fire air" of Carl Wilhelm Scheele in 1772,[19] a Brit published it first, the "dephlogisticated air" of Joseph Priestley in 1775[3] and a Frenchman understood it first, the "oxygen" of Lavoisier in 1775–1778.[2] Recent studies[20] suggest that Lavoisier may have heard about Scheele's and/or Priestley's work directly or indirectly: But, there's surely credit enough for all three to split the "retro-Nobel" awarded by Carl Djerassi and Roald Hoffmann in their sparkling play, *Oxygen*.[1]

At the Arsenal, Lavoisier extended his work on oxygen (his name from the Greek for "acid-forming") as the mediator of fire and life. He also proposed the first systematic enumeration of elements, a precursor of Mendeleev's periodic table.

> *Thus, while I thought myself employed only in forming a No-*
> *menclature, and while I proposed to myself nothing more than*
> *to improve the chemical language, my work transformed itself*
> *by degrees, without my being able to prevent it, into a treatise*
> *upon the Elements of Chemistry.*[13]

The treatise helped him to formulate the law of the conservation of matter: Nothing is lost in a chemical reaction. In perhaps his most vital work, he showed that living beings transform oxygen in the course of respiration, that they consume energy and generate heat and that muscular exercise burns calories as a candle does. One can measure this process. He called it calorimetry.[1]

Alas, Lavoisier fell to the guillotine, not for his science but his business interests. Lavoisier was one of approximately two dozen partners—among them, Mme. Lavoisier's father—in a private, for-profit corporation, the Ferme générale. The Ferme functioned as a crop-inspection and tax-collecting agency working on behalf of the crown. The proceeds from Lavoisier's share of Ferme générale revenue are said to have paid for his experiments at the Arsenal.[17]

After July 14, 1789, Lavoisier became a staunch supporter of the liberal constitutional monarchy set up in response to the Bastille uprising. In the upbeat interregnum of 1790, he wrote to Benjamin Franklin of the two revolutions that were still in progress. Lavoisier, supported by his colleagues at the Academy of Sciences, had overthrown the reigning "phlogiston" theory. He exulted to Franklin: "Here then a revolution has taken place in an important part of human knowledge since your departure from Europe." He was equally happy with the change in French political life: "After having brought you up to date on what is going on in chemistry, it would be well to speak to you about our political revolution. We regard it as done and without possibility of return to the old order."[21]

**Radical Terror**

The old order didn't return, but the Reign of Terror descended, and the tumbrels rolled. By 1792, the radical Jacobins were calling for heads, not only of the body politic. In quieter times, Jean-Paul Marat, that self-proclaimed genius of "optics, soap bubbles and medical electricity," had been denied admission repeatedly to the Academy. Unleashed by the Terror, he called for abolition of all of the academies. In his broadsheet, *L'Ami du Peuple*, he attacked Lavoisier personally:

*I denounce . . . the leader of the chorus of charlatans, Sieur Lavoisier, son of a land-grabber, apprentice-chemist, pupil of the Genevan stock-jobber [Jacques Necker, Louis XVI's finanace minister], a Fermier-géneral, Commissioner for Gunpowder and Saltpeter, Governor of the Discount Bank, Secretary to the King, Member of the Academy of Sciences.*[22]

Together with the constitutional monarchy, the Ferme générale was swept aside in the course of the Revolution. The radical Jacobins moved from mass protest to mass murder in what came to be called the "September massacres" of 1792. Lavoisier's laboratory at the Arsenal was shut forever. The Academy of Sciences dissolved shortly thereafter. In November 1793, Lavoisier, his father-in-law and 26 others of the Ferme générale were imprisoned and accused of "having plundered the people and the treasury of France, and of having adulterated the nation's tobacco with water, etc." They were each found guilty after a one-day trial and condemned to the guillotine on May 8, 1794.[22]

Two apocryphal quotations survive from the trial. One is attributed to Jean-Baptiste Coffinhal du Bail, the judge who dismissed appeals that cited Lavoisier's scholarly contributions to chemistry and the nation: "*La Republique n'a pas besoin de savans ni de chymistes!* [The Republic needs neither scholars nor chemists!]"[23] The other, attributed to the mathematician Joseph-Louis Lagrange, may be more authentic:

*It took them only a moment to sever that head, and a hundred years perhaps will not suffice to produce another like it.*[24]

Close enough: Albert Einstein was born in 1879.

# Experimental Errors: Paul Bert and the Alabama Tenure Killings

**PROFESSOR SAID TO BE CHARGED AFTER 3 ARE KILLED IN ALABAMA** . . . *Three faculty members at the University of Alabama in Huntsville were shot to death, and three other people were seriously wounded at a biology faculty meeting on Friday afternoon . . . a biology professor, identified as Amy Bishop had applied for tenure, been turned down, and appealed the decision. She learned on Friday that she had been denied once again.*

— *The New York Times*, 2010[1]

*The effects of lowering or raising the barometric pressure can be countered by increasing or decreasing the $O_2$ fraction in the air. Extreme hyperoxia modifies cellular metabolism of all living beings: this is $O_2$ poisoning, the Paul Bert effect.*

—P. Dejours and S. Dejours, 1992[2]

**FATAL BALLOONING: THE SAD STORY OF THE ZENITH**
*. . . At 7000 metres they lost consciousness of their acts, and were almost powerless to raise a hand even to take hold of the respirator. Sivel did so, and inhaled large doses of oxygen and appeared to have become excited unduly thereby. [He] went on cutting down bags of sand. "Yes," he cried gaily, "and happy the one of us that returns!" When the balloon had descended to 6,000 metres Tissandier came to himself to find his companions dead and the Zenith falling like a stone.*

— *The New York Times*, 1875[3]

It's NOT OFTEN THAT THE FATAL ACTS of experimental biologists have commanded banner headlines in *The New York Times*. The latest was Amy Bishop's rampage on February 12, 2010, at the University of Alabama, Huntsville, and the earliest was the Zenith balloon disaster of April 18, 1875, with which the respiratory physiologist Paul Bert was associated. And although the two events clearly differ in time and place, they both have roots in national policies. The aeronauts of the *Zenith* died in the quest for French mastery of the air; the killings in Huntsville showed that we lack adequate gun control in America. Read back-to-back, the stories have two features in common: both principals were biologists who had studied a respiratory burst, and both mortal mishaps were blamed on mental aberration.

### This Is Not a Whodunit

On February 13, *The New York Times* reported that Dr. Amy Bishop, a 45-year-old assistant professor of biology at the University of Alabama, Huntsville, had been arraigned the day before for the shooting of six fellow faculty members: "The shootings took place after Dr. Bishop learned that she had lost her long battle to gain academic tenure at the university."[1]

Bishop had written her 1992 PhD thesis at Harvard University on the "respiratory burst" of phagocytes and neurons; papers regarding this work appeared sporadically until 1998.[4,5] The point at issue in Huntsville was her contribution to research since her arrival at Alabama in 2003. Colleagues reported that Bishop became severely upset after tenure was denied and vented her fury in gunplay. The weapon chosen was a 9-mm semi-automatic pistol with 12 rounds in the chamber.[6]

An eyewitness account of the mayhem in Huntsville was given by Dr. Debra Moriarity, dean of graduate studies. Thirteen people had crowded into a small conference room to discuss routine departmental matters. For one hour, Amy Bishop sat quietly in a corner, and then

> *Moriarity heard the first crack of a pistol. She looked up to see the shooting of Dr. Maria Ragland Davis. Moriarity dove under the table, crawled toward Bishop, and grabbed onto her legs. Bishop shook free. Moriarity kept crawling, but Bishop pursued her out of the conference room. Moriarity pleaded for her life, but Bishop pulled the trigger. The gun clicked, and clicked again, Moriarity said. She crawled past Bishop and back into the conference room, where survivors*

*quickly barricaded the door with a coffee table and a small refrigerator.*[6]

Gopi Krishna Podila, the department chair, died from a single shot to the chest; Dr. Adriel Johnson and Dr. Davis, each died of a single shot to the side of the head. Three others were wounded, two critically.

Following Bishop's arraignment on three counts of capital and three of attempted murder, her defenders sought motives for the mayhem. Her lawyer told reporters, "This is not a whodunit. This lady has committed this offense or offenses in front of the world. It gets to be a question in my mind of her mental capacity at the time, or her mental state at the time that these acts were committed."[7] Others were a touch more parochial. A Boston criminologist explained: "Being denied tenure when you're in your mid-40s at an out-of-the-way obscure rural campus in the deep South is a catastrophic loss, and people don't understand that."[8] But, the tenure-as-excuse argument seems tenuous.

Bishop had been implicated in earlier acts of gunpowder violence in Massachusetts. At age 20 in Braintree, she'd fired off three rounds from a shotgun, one of which killed her teenage brother after a "domestic argument." She and her husband had also been suspected of mailing pipe bombs to a Harvard professor in Newton, who had been less than thrilled with Bishop's postdoctoral work on the respiratory burst of neurons.[9]

**Publish or Perish**

The news stories didn't delve much into the science she had or hadn't been doing, so I followed up on the story on the web. To update another Harvard professor, Henry Wadsworth Longfellow,[10]

> *Lives of great men all remind us*
> *We can make our lives sublime,*
> *And, departing, leave behind us*
> *Websites on the sands of time . . .*

By mid-March, Amy Bishop's official bibliography had been scrubbed from the several websites on which it had been featured. Accessed on February 18,[4,11] it listed a total of 19 publications: Ten are original papers in refereed journals such as *Environmental Health Perspectives, Free Radical Biology & Medicine* and the *Journal of Neurochemistry*; nine are reviews and/or conference literature. *PubMed*, which cannot be scrubbed, lists 11, of which six are original research reports. Almost all of these describe work in Boston with her distinguished and now-deceased mentors, Paul Gallop and Manfred Karnovsky.[12] Reckoning that cells undergoing a respiratory burst must defend themselves from superoxide anion or NO, they identified a molecular protectant, pyrroloquinoline quinone (PQQ). As have others before and since,[13] they suggested that PQQ was a new vitamin.[4] Bishop's postdoctoral work in Boston sent her from lab to lab to emerge with modest additions to related literature.[14]

Altogether, the websites suggest that Dr. Bishop's later work in Alabama left few footprints on the sands of time. Her latest e-publication is a footnote to that judgment. Accessed early in March, the complaisant website of the *International Journal of General Medicine* (published by Dove Medical Press in New Zealand) listed a May 2009 article by Lily B. Anderson, Phaedra B. Anderson, Thea B. Anderson, Amy Bishop, et al., "Effects of selective serotonin reuptake inhibitors on motor neuron survival." Instead of the usual abstract, the journal noted:

> *Pending further investigation into alleged breaches of our manuscript submission criteria this paper has been removed from the dovepress.com website.*[15]

Amy Bishop and her husband James Anderson had listed three of their teenage children as coauthors. The youngest of the teens is 14; the oldest, 18-year-old Lily Bishop Anderson, is a student at the University of Alabama in Huntsville. The children's "institutional affiliation" was their home address.[16]

The results of this website quest suggest that after a respiratory burst of serious science in Boston, Amy Bishop's career was like an inflated balloon: untethered, adrift and heading for trouble.

## The Paul Bert Effect

The earliest *Times* account of a fatal error by an experimental biologist describes a real balloon, untethered, adrift and heading for trouble. On April 18, 1875, aeronauts Joseph Croce-Spinelli and Théodore Sivel perished while experimenting with oxygen at near-record altitudes aboard the *Zenith*. A third balloonist, Gaston Tissandier, barely survived.[3]

The voyage of the *Zenith* was an episode in the late 19th-century European race for space. Mastery of flight had become important in the Franco-Prussian war of 1870, in which balloons proved essential for observation, mail delivery and evacuation. Height meant mastery. In 1862, British balloonists had already ascended close to 9,000 meters, but in thin air, without oxygen, they developed cerebral symptoms and transient limb paralysis.[17] Meanwhile, German balloonists were busy developing high-flying dirigibles that became precursors of the Zeppelins.[18]

Determined to outdo these rivals, French science called on Paul Bert, who had succeeded Claude Bernard as dean of French physiologists. To prepare for the ascent of the *Zenith*, Bert had the aeronauts train for weeks in his new hypobaric bell-jar chamber. They gasped in the hypoxic air of simulated high altitudes and survived in the rush of pure oxygen—the respiratory burst.[19]

The *Zenith* took off at daybreak on the 18th of April, with Tissandier's log recording the mission in graphic detail: "The ascension from the gas-works at La Villette was accomplished favorably," and by one p.m., they were comfortably cruising at 4,600 meters (15,000 feet).[20] Bert had provided the crew with bags of oxygen, but his calculations fell short of the amount needed. As the *Zenith* reached 7,000 meters (22,960 feet), Tissandier felt "weak all over." He saw the others become pale, and he took a small whiff of oxygen. Sivel soon dropped bag after bag of ballast, and the balloon rose rapidly. Tissandier became so weak that he was unable to hold up his head. Still lucid, he again reached for the oxygen tube but couldn't lift his hand to grasp it. At 8,000 meters (26,240 feet), he passed out for about 20 minutes "Reviving for a moment, he was shaken by the arms, by Croce-Spinelli, who told him to throw out ballast, for the balloon was descending at a very rapid rate."[21] Sivel had also revived; he took one long last whiff of pure oxygen. Exhilarated by the gas and mentally confused by oxygen toxicity at that low barometric pressure, Sivel dropped more and more ballast: "Yes," he cried gaily, "and happy the one of us that returns!"[3] The craft literally shot for the moon: By 9,000 meters, all three men became unconscious. Unmanned, the balloon dropped rapidly in altitude. Tissandier came to his senses again at 6,000 meters, only to find his colleagues dead. He gained control of the craft to bring the fatal voyage of the *Zenith* to an end.

It remains unclear to this day why there was too little oxygen onboard; Bert's is the only verbatim account, and it's not a little self-serving: "I was then absent from Paris and my letter [advising them to increase oxygen gradually during ascent] arrived too late . . . and they drew from my observations only this conclusion which was so fatal, that they should wait for absolute necessity to make use of the gas bags."[21] Those "fatal observations" may have killed two aeronauts, but Bert had learned that hyperoxia, the respiratory burst, can excite and then extinguish the mind.

Croce-Spinelli and Sivel lie under a graceful monument in the Père-Lachaise cemetery; Tissandier gained fame as editor-in-chief of *La Nature* (the Gallic answer to *Nature* of London); Paul Bert became minister of Education and Worship in Léon Gambetta's cabinet in 1881. Experimental biologists remember him for the Paul Bert effect (Sivel's oxygen toxicity), "caisson disease" and for an early treatise on anesthesia and tissue grafting. He also served as president of the Société de Biologie. French politicians honor him as a founder of the French Socialist Party and the country's educational system (*l'école gratuite, laïque et obligatoire*, i.e., free, secular and compulsory).[19] In 1886, Bert died of cholera in Hanoi, where he had been sent as governor-general.[22] American tourists in Paris recognize his name as that of a bistro near the Hôpital Saint-Antoine and a louche corner of the Paris flea market.

### Shoot, Shoot, Shoot

What can we conclude from these two public catastrophes, separated in space and time but linked by the respiratory burst and altered mentation? Paul Bert taught us to vary inspired oxygen inversely with the barometric pressure in-flight and beneath the sea. We hail him correctly as the father of aviation and submarine medicine. Thanks to his hypobaric chamber, we fly, fully sentient, in pressurized cabins at 30,000 feet as we read about fully manned submersibles going after deep-sea life 12,400 feet beneath the sea.

As for Amy Bishop? Whatever her mental state, we assume that the state of Alabama will grant her lifetime tenure in one or another state institutions. However, the bigger story is not of tenure denied. "There are 80 gazillion disgruntled junior faculty, postdocs, and grad students who don't go into a faculty meeting guns blazing."[23] Disgruntled folks, tenured or not, can't go into a meeting with guns blazing—or shoot their brother—unless they have guns. Guns are the problem, not tenure.

In keeping with the Second Amendment's "A well regulated Militia, being necessary to the security of a free State," Bishop knew how to bear arms. A few days before the killings, she had practiced with her 9-mm semi-automatic at an indoor shooting range. At that faculty meeting, "It was shoot, shoot, shoot, very regular, both hands on the gun," an observer reported. "She absolutely knew how to handle that gun."[6]

Across the pond, they are astonished: A Scots paper summed it up: "Every few weeks a mass shooting occurs which would scandalize any other developed nation. . . . On an average day, 85 Americans die from gunshot wounds—more than 30,000 each year."[24] It's time to stop this madness. Let's begin by keeping guns off campus and NOT teach kids about the NRA while skipping Thomas Jefferson.[25] To update Longfellow again:

> *Let us, then, be up and doing,*
> *With a heart for any fate;*
> *Still achieving, still pursuing,*
> *Learn to labor and not wait.*[10]

# Monumental Revolutions:
# Scientific, Sanitary and 'Omic

*The so-called "scientific revolution" of the sixteenth and seventeenth centuries outshines everything since the rise of Christianity and reduces the Renaissance and Reformation to the rank of mere episodes, mere internal displacements, within the system of medieval Christendom.*

—Herbert Butterfield, *The Origins of Modern Science*, 1949[1]

*I think that there is more barbarity in eating a man alive than in eating him dead; and in tearing by tortures and the rack a body still full of feeling . . . on the pretext of piety and religion, than in roasting and eating him after he is dead.*

—Montaigne, "Of Cannibals," ca. 1580[2]

*Let the land be covered with sanitary associations and it will soon stand as much a landmark for the health, happiness and comfort of its people, as it is now a beacon light for the politically oppressed of other lands.*

—Quarantine and Sanitary Convention USA, 1859[3]

*The genomic revolution is here—are you ready?*

—American Museum of Natural History, 2001[4]

## 'Ome Depot

To ANSWER OUR COLLEAGUES at the American Museum of Natural History, we're not only ready for the genomic revolution, we're smack in the middle of the next one. I realized this when I clicked on a word new to me in the

103

program of the 2009 Ubiquitin Discovery Conference: "Is it likely that the ubiquitome will replace the kinome as the most promising source of new medicines?"[5] I got it. Neologism or not, "ubiquitome" seemed as straight-forward as "kinome," or for that matter "proteome" or "metabolome," each of which pops up often nowadays. The corresponding research field of each is called "omics" in keeping with "genomics." So here we are, deep into the 'omic revolution; sure enough, there's the corresponding journal *Omics*, founded in 1995. The journal dutifully defines its subject:

> *Following the success of the human genome project effort, several other "omic" disciplines have emerged, with the goal of analyzing the components of a living organism in its entirety.*[6]

Folks in the 'omic community are engaged in mapping out not only the complete set of proteins produced in a cell (proteomics) but also the complete set of altered methylations in the genome (epigenomics), the complete set of mutational phenotypes (phenomics), the complete set of organic small molecules (ligandomics) *und so weiter*.[6] One lapses easily into German because "genome" was coined by the German botanist Hans Winkler (1877–1945). Winkler called the sum total of genes in a set "genom" (1930) taking Johannsen's name for the genetic material, the "gene" (1905) and joining it to the tail end of Waldeyer-Hartz's name for stuff in the nucleus that took up basic dyes, the "chromosome" (1888). One might say that the 'omic revolution started out a century ago in a hunt for the complete set of heritable polyanions.

The etymology comes from a cheerful comment written by Joshua Lederberg in 2001 titled "'Ome Sweet 'Omics" in which Josh and linguist Alexa McCray sanctified the 'omic revolution:

> *Genomics and Proteomics are the buzzwords of the dawning millennium. There is no counting of "omic" sites to be found on the Web. That most of these terms, old and new, have been contrived as slogans to attract attention, does not diminish their likely substance, and they are embedded in the advancing edge of science and technology.*[7]

'Omics has certainly attracted attention in fields ranging from anergy to zoonoses, and their promise of advancing the edge of science has also been kept. For example, together with its handmaiden of systems biology, 'omics has permitted broad-scale mapping of mutational patterns in cancer by Arnold Levine's group,[8] deciphering of complex gene regulatory networks by Eric Davidson,[9] and Raul Rabadan's analysis of the H1N1 swine flu virus.[10]

But the 'omic revolution has not just given us new facts: it has changed the way biologists think. Pioneers of 'omics and systems biology claim that they have either overturned traditional, hypothesis-driven research completely or at the very least found an alternate way to do science.[11,12] The novel techniques of microarray analysis, of "connectivity" or "molecular interaction" mapping, of kinetic simulation of cell processes have been made possible by information technologies that owe as much to Oracle and SAP as to Krebs and Chargaff. As discovery research becomes replaced by information-mining, data no longer lead to hypothesis, but make hypothesis unnecessary.[12]

'Omics, systems biology and computer simulation promise to give biomedical science the power of visiting sites not seen before because the route has been mapped in silico. Virtual or real, there's no need to worry over cause and effect: effect is the cause. Glenn Evans wrote at the beginning of the 'omic revolution:

> *When an "omic" experiment is completed, the data could be entered in the simulation and the model "run" . . . Ultimately, based on a more profound understanding of gene networks and systems rules than currently available, the simulation space would allow construction of an* in silico *mammal.*[13]

*In silico veritas!* The chip is what has made both 'omics and systems biology possible. It poses a challenge to the tradition of inductive, hypothesis-driven research that has dominated science since the dawn of the scientific revolution.

### The Scientific Revolution: Crème Brûlée

Modern historians of science are of several minds on the subject of scientific revolutions. Herbert Butterfield (1900–1979) of Cambridge believed that there was one Scientific Revolution and that it changed the world.[14] Harvard's I. Bernard Cohen (1914–2003) taught that there were many revolutions in science—led by Copernicus, Lavoisier, Darwin etc., and that they all changed the world.[15] Thomas Kuhn (1922–1996) traded the notion of "revolution" for a "paradigm shift" that arises when new, empirical facts become incompatible with widely held beliefs. He studied several such events and argued that some have changed the world, others not.[16] No one disputes that the scientific revolution(s) started in Renaissance Europe.

Butterfield dates the first Scientific Revolution to 1543, the year Nicolaus Copernicus published *De Revolutionibus Orbium Coelestium* in Nuremberg and Andreas Vesalius published *De Humani Corporis Fabrica* in Venice.[14] The astronomer described the correct orbit of the earth around the sun; the anatomist drew the first correct map of the human body. Both used empirical observation to correct medieval notions of heaven and man, posing a major challenge to the established church. The reaction was swift and brutal. Dissenters, or imagined dissenters, were subject to harsh detainment, torture and burning at the stake. A later critic noted the era: "The ferocious theologians are after them with a *Bible* in one hand and a fiery fagot in the other."[17] Between 1546 and 1600, three Renaissance luminaries who made the scientific revolution possible were burned at the stake: Étienne Dolet in Paris (1546), Michel Servetus in Geneva (1553) and Giordano Bruno in Rome (1600). The best of the best of a generation were burned, or, as Ann Weissmann has quipped, *crème de la crème brûlée.*

Most scientists know the Servetus story. An astronomer, polymath and translator of Ptolemy, Servetus was the first to discover the pulmonary circulation in man: his natural philosophy ran afoul of John Calvin. Many also know the tale of Giordano Bruno, the rationalist philosopher, astronomer and early champion of Copernicus. His description of medieval cosmology could be used by a skeptical referee of a new submission to a modern medical journal: "*Se non è vero, è molto ben trovato* [If it's not true, it's very well invented]."[18] The new science of Servetus and Bruno shook the pillars of the church:

> It must have seemed a real blow to man's pride to be told that
> his planet had been shifted from a fixed place at the center,
> only to become just "another planet" as Copernicus called it
> and physically a rather insignificant planet at that.[19]

Chroniclers of science tend to bypass Étienne Dolet, nowadays chiefly remembered as a humanist, printer of Rabelais's *Gargantua* and author of an early essay on translation.[20] But recall that the first generation of Renaissance scholars called themselves humanists rather than scientists. Humanist ideas were based on rediscovery of the original texts of antiquity that had survived the dark ages of Christendom in libraries of the Arab world. The works of Aristotle and Plato, Ptolemy and Euclid arrived in Europe in need of reinterpretation. As in the 'omic revolution of today, transcription, translation and digital reproduction spurred the scientific revolution. Movable type, which Gutenberg had introduced 1439, played the role of the microchip. Translators wrote the program, printers provided the software and the stage was set for 1543.

But woe to translator or printer who crossed the line. Although Étienne Dolet was tortured, strangled and burned on the Place Maubert for heresy, the only "crime" of which the Sorbonne clerics found him guilty was that of amending an approved version of Plato's Dialogue *Axiochus*. Translating from the original Greek into colloquial French, he added one phrase, *rien du tout* (nothing at all) to a passage that read "and when you have died, death will also not be able to do anything, since you will no longer be. Nothing at all."[20]

The elongation—a slap in the face to believers in an afterlife—was in keeping with his principles of translation in a classic book (*La manière de bien traduire d'une langue en aultre*, 1540). This acerb work has outlived his ashes: "While translating, you must not be enslaved to the extent of rendering word for word. And if anyone does so, this comes from his impoverishment and deficiency of wit."[21] But clerics had the last word, the printer whose wit cleared the path to the modern world was set aflame in August of 1546.

### Twice Burned

It took epidemic cholera, the urban renewal plans of Baron Haussmann and the Sanitary Revolution to revisit the martyrdom of Étienne Dolet. The Place Maubert, where Dolet was immolated, had over the centuries been a raffish venue for impoverished students from lycées and the Sorbonne, for street artists, vagabonds and *clochards*.[22] It was also situated at a confluence of hovel-lined streets that ran open sewage from the colleges and seminaries to the Seine by way of the fetid Bièvre Canal. But, in response to the waves of cholera that swept Paris in 1832 and 1865, and after it became clear that cholera came from contaminated water, the map— and smell—of Paris changed forever. Baron Georges-Eugène Haussmann, prefect of the Department of the Seine from 1853 to 1870, presided over the creation of parks, boulevards and an urban plan that replaced quaint squalor with middle-class solidity. He directed construction of a massive sewer system and of artesian wells to provide clean water. The cloaca of the Bièvre Canal was paved over, and the wide, open Boulevard Saint-Germain brought sun and light.[23] As a result, when the progressive Third Republic came to govern France, Paris was spared during the cholera epidemics of the 1880s that scourged Europe from Hamburg to Genoa.

The legacy of the Third Republic (1870–1940) is not only social democracy but also *laïcité*, the separation of church and state in France. No surprise then that, in 1889, the government erected a monument to Étienne Dolet on the very spot he was burned. Over four decades the Dolet monument became a gathering place for union marches, feminist rallies and left-wing protests. As Margaret Cohen recalls, it prompted memoirs, real and surreal.[22] All that changed under German occupation, when Nazi sympathizers saw to it that Dolet's statue was among the first of Gallic heroes to be melted down.

Dolet's fellow incendiary, Giordano Bruno, also had a statue dedicated to him in 1889 by a progressive Italian government that wrested Rome from papal rule. Bruno's statue still presides over the Campo dei Fiori of Rome. A tourist circular notes: "The anniversary of Bruno's death is celebrated by the Italian Association for Freethinking. . . . In today's Italian political environment the monument and the ceremony are an embarrassment to politicians who fear to displease the Catholic hierarchy."[24] So much for political progress and the Scientific Revolution.

## The Sanitarian Angel

Political progress and the Sanitary Revolution may have had equally rough times here in America, but their monuments remain. Progress and clean water are commemorated by two statues in New York's Central Park: Augustus Saint-Gaudens' equestrian figure of General William Tecumseh Sherman (1899) towers over the plaza at the southeastern corner, while Emma Stebbins's *Angel of the Waters* (1873) tops the Bethesda Fountain near its center. The imposing gilt general, preceded by a glorious Victory sports a cloak and a sword. Wags claim that well-draped Victory is pointing the way to Bergdorf Goodman. Stebbins's angel, lily in hand, shades the fount of healing with her hovering wings. She sheds the blessing of Union on the water below, supported by four cherubs representing Health, Purity, Peace and Temperance.[25] Happily the fourth cherub was unknown to Generals Grant or Sherman.

As with Étienne Dolet, we owe the Central Park monuments to an urban planner and an epidemic disease: Frederick Law Olmsted and (again) cholera. Saint-Gaudens and Stebbins celebrate abolition and sanitation; or reform and chloroform, as Dr. Oliver Wendell Holmes had it.[26] Reform and chloroform were also the driving force behind the U.S. Sanitary Commission, the executive secretary of which was Olmsted, an arch abolitionist, journalist and landscape architect. Olmsted's greatest architectural achievement was the master plan for Central Park, which he and Calvert Vaux drafted in 1856.

William Cullen Bryant, the poet, had launched the campaign for Central Park in the 1840s; but many argue that the park was actually built in response to the cholera epidemics of 1848–1849; its construction was a major issue of the 1850 mayoral campaign.[27] Cholera had arrived in full force in the spring of 1849, following a year of revolution and large-scale emigration from Europe. But despite strict quarantine, inspections of new immigrants at the port and a great bout of street scrubbing, the summer of 1849 brought 18,000 cases of cholera and 8,000 deaths in a city of half a million inhabitants. At its high-water mark, cholera killed 1,178 New Yorkers in one day: August 11, 1849.[27]

Not until the winter did the epidemic abate, and by then clean water and open spaces had become a political issue. It was clear that land free of "squatters, pigs and shanties" was needed. The cost of Central Park was staggering at the time—$5.5 million for the land alone. But proponents for the park argued that it was a price worth paying. Uncontaminated land was needed on which to place the reservoirs needed while the water aqueducts in Croton, New York, were slowly put in service. But there was more to the park than water. The local commissioners charged with construction of Central Park hoped that this "barren waste might be converted into a good healthy pair of lungs for the entire population of the most enterprising City in the World."[25]

When cholera struck again at the end of the Civil War, in 1866, the park had been largely completed. *The New York Times* for August 19, 1866, lists the bills of mortality for cholera, then raging in the city. However, in the immediately adjacent column, its editors proudly note that Mr. Olmsted's Central Park was ready for summer recreation, having been won from "a dreary waste of sterile rocks, rising boldly and defiantly in the face of thrift and enterprise, relieved now and then by filthy sink-holes and pools of stagnant water . . . an excrescence on the fair features of the city." The city was therefore ready to respond to cholera when the Sanitary Commissioner George Templeton Strong decreed that "Sanitation, not quarantine is the answer!" He was correct, and when a cholera alert was sounded in April of that year, the city used both approaches, quarantine and sanitation. As a result, although New York had grown to a population of 750,000 when the 1866 epidemic struck, only 1,200 of its citizens contracted cholera and only

600 died.[28] The sanitary revolution had done its job, in New York as it had in Paris. The Angel was unveiled three years before Robert Koch launched the bacteriological revolution with his discovery of the znthrax bacillus.

## 'Omics and the Angel

Emma Stebbins (1815–1882) sculptor of the *Angel of the Waters* fountain, is an important figure for several traditions—the genteel, the feminist, the lesbian, the fine arts. Brought up in a well-off New York family—her brother was a commissioner of Central Park—she went off to Rome to study sculpture. There she met another American, the famous actress Charlotte Saunders Cushman (1816–1876): they found each other on a trip to Naples and determined to live together ever after. Sadly, shortly after the *Angel of the Waters* fountain was unveiled in 1873, Cushman died of breast cancer. Stebbins faithfully tended her dying mate, and ill herself, sank into depression, never completing another sculpture.[29] She did, however, compose a biography of her famous mate, featured in *The New York Times* of May 26, 1878, as a full-cap review:

> THE SUCCESSOR OF SIDDONS. CHARLOTTE CUSHMAN'S LETTERS. THE PITH OF MISS. STEBBINS' BIOGRAPHY A FORTHCOMING VOLUME OF RARE INTEREST CAREER OF AMERICA'S GREATEST ACTRESS A STORY OF GENIUS AND PERSEVERANCE. HER FIRST APPEARANCE IN DRAMA. HER FIRST VISIT TO EUROPE. HER LATER CAREER.[30]

Misfortune continued as Stebbins withered away in 1882 from what was then called phthisis—that wasting systemic response to tuberculosis, effects we now attribute to TNF and related cytokines. It was the same year that Robert Koch discovered the tubercle bacillus: a year later, the cholera *vibrio* was found.

I'd bet that right now, today, there are data that spell out a cure for those who are dying of diseases as close to scientific solution as was the malady that killed Stebbins at the dawn of the bacteriological revolution. I'll bet the data lurk somewhere *in silico*, in those glowing, growing connectivity maps, hidden in base-pair mines, in cyberspace, in websites devoted to genomics, pharmacogenomics, ubiquitomics and so on. Some or any of those 'omics will surely lead us to that Angel of Healing. That's my hypothesis.

# Quorum Sensing on the Airbus Wing

*Scrambling for the exits and carrying the helpless, they perched ankle-and then knee-deep atop the wings as an improvised armada of tour boats and ferries streamed to their rescue . . . witnesses described a scene of level-headed teamwork to rescue the weak and infirm.*

— *The Washington Post*, January 16, 2009[1]

*We at once see that those animals which acquire habits of mutual aid are undoubtedly the fittest . . . we may safely say that mutual aid is as much a law of animal life as mutual struggle but that, as a factor of evolution it most probably has a far greater importance.*

—Prince Pyotr Kropotkin, 1902[2]

*Bacteria communicate extensively with each other and employ a communal approach to facilitate survival in hostile environments.*

—Shadaba Asad and Steven M. Opal, 2008[3]

## Mutual Aid and Quorum Sensing

THE MASTERFUL DITCHING in the Hudson River on January 15, 2009, of US Airways Flight 1549 and the rescue of all its 155 passengers provided high drama on a cold winter day. That "Miracle on the Hudson" preoccupied national television from the start and produced a real-life aviation hero, Captain Chesley "Sully" Sullenberger. It also evoked a behavioral response that a century ago was called "mutual aid" by Prince Kropotkin[2] and nowadays would be described "quorum-sensing" by microbiologists.[3]

Within minutes after the plane alighted safely on the frigid Hudson, hatches opened and passengers, guided by attendants, emerged in orderly fashion through the emergency exits. "Some passengers began to wail, but witnesses described a scene of level-headed teamwork to rescue the weak and infirm."[1] The Airbus remained adrift and partially submerged in the ice-cold water. Groups of passengers spread out on both wings of the jetliner; then as dry space on the wings gave out, a few moved to rafts. Within minutes, the drifting Airbus was surrounded by ferry and sightseeing boats, then by police vessels, fireboats, tugboats and Coast Guard craft. Their crews tossed life rings and pulled the passengers to safety, their prows held the plane in place against the current. Helicopters brought wet-suited police divers, who dropped into the water to help with the rescue.

We'd call that kind of response mutual aid, for sure. But, what about that optimal clustering of folk on the teetering wing, that biofilm of rescue rigs in the water, that bobbing flotilla of vessels securing the plane, the airways abuzz with signals of distress and rescue? Why that sounds a lot like "quorum sensing,"[4] an evolutionary strategy used by microbes to effect a "coordinated population response . . . explaining both cooperation and communication."[5]

## Let Them Be Sea Captains if They Will

Another theme emerged: among the first ferry captains to arrive at the scene was Brittany Catanzaro, the 20-year-old captain of the New York Waterways ferry *Thomas Kean*. It occurred to me that the drama of the Airbus on the Hudson was a *prima facie* argument for the equal rights amendment. Brittany Catanzaro and her crew plucked 24 passengers from the ditched flight to safety. "I've been on the water since I was two years old," the captain told *The New York Times*. "I pulled out of Pier 79, I looked for any kind of southbound traffic, and I saw the plane there. . . . It was hard to stay next to it, but you practice that by throwing life rings in the water and trying to stay alongside them."[6]

The pioneering American feminist Margaret Fuller would have been proud. Fuller, who served on Horace Greely's *New York Tribune* as America's first female foreign correspondent, argued for women's equality in the moral, social and occupational spheres. Her 1845 work, *Woman in the Nineteenth Century*, anticipated Captain Catanzaro's feat by over a century and a half:

> But if you ask me what offices they may fill, I reply—any. I
> do not care what case you put; let them be sea captains if they
> will. I do not doubt there are women fitted for such an office.[7]

Neither would Fuller have doubted the fitness of Sheila Dale, Donna Dent and Doreen Welsh, the three veteran flight attendants who safely guided the passengers from seat to hatch to wing to shore. Each was over

50 years old and retained her job thanks to legal action taken by Eulalie Cooper, a flight attendant who in the sex-kitten days of the 1960s was fired by Delta Air Lines for being married.[8,9]

Fuller ranked among our nation's leading literary figures in the antebellum period; she had edited *The Dial* with Ralph Waldo Emerson and became a cultural critic for the *New York Tribune*.[10] She shared the meliorist convictions of the Brooke Farm transcendentalists who, half a century before Prince Kropotkin, believed that mutual aid was a human obligation:

> *With simultaneous vibration the hearts of all their circle acknowledged the divine obligation of love and mutual aid between human beings. Food, clothing, medicine were all offered freely.*[11]

In the course of her foreign adventures, she married the Marquis Ossoli, an Italian patriot who fought bravely on the barricades for the Roman Republic against the royalist-papal armies of France and Austria during the Revolution of 1848–1849. Fuller freed herself to work day and night as a volunteer nurse in the Tiber Island hospitals. In those bloody field stations, as a posthumous account had it, "the weather was intensely hot; her health was feeble and delicate; the dead and dying were around her in every stage of pain."[12]

Ironically, Fuller died less than five years after her sea captain plea, because of the unfitness of a male sea captain in waters around New York. On July 19, 1850, Margaret Fuller, her husband and infant son were drowned in a shipwreck on her return voyage from the Italian wars. An inexperienced first mate had run the merchantman *Elizabeth* aground on a sandbar off Fire Island. Her body was never recovered. Days later, her old friend Henry David Thoreau wandered the sands looking for remnants of the Ossolis, but nothing had washed ashore.[10] If only the *Elizabeth* had been tended to by a sea captain as professional as Brittany Catanzaro!

### An Anarchist Editor of Nature

Prince Pyotr Kropotkin (1842–1921), who fathered the principle of mutual aid, was no mean feminist himself. He had taunted his tsar, Alexander II, who in turn had him imprisoned. The prince escaped and found refuge in England:

> *In spite of the open hatred of Alexander II for educated women,—when he met in his walks a girl wearing spectacles and*

*a round Garibaldian cap, he began to tremble, thinking that*
*she must be a nihilist bent on shooting at him; in spite of*
*the bitter opposition of the state police, who represented every*
*woman student as a revolutionist; in spite of the thunders and*
*the vile accusations . . . there are now in Russia more than six*
*hundred and seventy women practicing as physicians.*[13]

The prince, a natural scientist who had served as secretary of the Russian Geographical Society, arrived in London in 1886 and soon called at the offices of *Nature*, his favorite journal. He was "most cordially received" by a sub-editor, Mr. J. Scott Keltie, who was looking to increase the journal's international coverage. Kropotkin, working under the exile's pseudonym of Levashóff, showed the approving Keltie a few samples of his writing and soon the anarchist prince was given a table in the office of *Nature* and asked to review foreign publications and books. The multilingual prince performed these editorial duties flawlessly until "One day, however, Mr. Keltie took from the shelves several Russian books, asking me to review them for *Nature*. I looked at the books, and, to my embarrassment, saw that they were my own works on the 'Glacial Period' and the 'Orography of Asia.'"[14]

Kropotkin confessed that he could not comply with this request and admitted the works were his own. His cover was blown, and the prince retired from *Nature*. He went on to edit the British anarchist magazine *Freedom*, the motto of which remains a tribute to his politics—and his evolutionary theory: "Anarchists work towards a society of mutual aid and voluntary co-operation."[15] In his *Memoirs of a Revolutionist* (1899), the prince complained that the Darwinian struggle for existence had been misinterpreted in the biological as well as the social realm. He cried that "there is no infamy in civilized society, or in the relations of the whites towards the so-called lower races that has not been excused by the Darwinian 'struggle for existence.'"[16]

Kropotkin—who would have been thrilled by that Airbus on the Hudson—was the first to draw attention to "mutualism" in animal behavior, a notion based on his field observations in Siberia. For him the rabbit was the very model of the social animal; its persistence—and capacity for mass production—was an argument against the dog-eat-dog world of popular (social) Darwinism.

The rabbit, he told his friend Ford Madox Ford, stood out against the fiercer aspects of selection. Defenseless and adapted for nothing in particular, it had outlived "the pterodactyl, the Hyrcanian tiger and the lion of Numidia."[17] One wonders how he would have described those Canada geese that brought down the Airbus. He'd probably bet that the Canada goose will go on to outlive the Airbus—and Boeing for that matter—as it has outlived the Concorde, Lockheed and the DC-3.

Anticipating E. O. Wilson, Kropotkin proposed mutual aid as a force in natural selection which ensured that group survival need not depend a Hobbesian struggle between the weak and the strong. The prince was neither a fan of intelligent design nor an eco-sentimentalist: if mutual aid gave a species selective advantage, it had nothing to do with "love":

> It is not love, and not even sympathy (understood in its proper sense) which induces a herd of ruminants or of horses to form a ring in order to resist an attack of wolves; not love which induces wolves to form a pack for hunting; not love which induces kittens or lambs to play, or a dozen of species of young birds to spend their days together in the autumn; and it is neither love nor personal sympathy which induces many thousand fallow-deer scattered over a territory as large as France to form into a score of separate herds, all marching towards a given spot, in order to cross there a river. It is a feeling infinitely wider than love or personal sympathy—an instinct that has been slowly developed among animals and men in the course of an extremely long evolution, and which has taught animals and men alike the force they can borrow from the practice of mutual aid and support, and the joys they can find in social life.[18]

H. W. Bates, the father of behavioral genetics, called Kropotkin the truest Darwinist: mutual aid was a law of nature on a par with Darwin's "group selection."[16] But in those days before the gene and DNA, neither Kropotkin nor Darwin had a clue as to the biological underpinnings of fitness or natural selection. How could generations of wild horses know exactly how many of them were needed to form a circle against wolves? How exactly did an "extremely long evolution" work out the optimum number of Canada geese required to fly in almost radar-perfect V-formation?[19] Nowadays we turn to molecular and behavioral genetics for answers to such questions, some of which will likely be based on what we've learned of "quorum sensing" in microbes.

## The Vibrio, the Squid and the Quorum

"Quorum sensing" was introduced in 1994 by Fuqua et al. to explain mutualism between a bioluminescent marine bacterium, *Vibrio fischeri,* and the Hawaiian bobtail squid, *Euprymna scolopes.*[4] Both benefit from cohabitation: the bacteria are nourished by nutrients from the squid, and the squid gains a unique mating advantage from the vibrio. From early in their development, the squid take up the vibrios, which become concentrated in light organs distributed along the body of the cephalopod.[20] When a sufficient number of vibrios accumulate in these organs, the colony senses that it's time to generate light by turning on luciferase genes:

> . . . *certain bacterial behaviors can be performed efficiently only by a sufficiently large population of bacteria. We describe this minimum behavioral unit as a quorum of bacteria.*[4]

Mutualism between vibrio and squid serves to generate what J. W. (Woody) Hastings, of Harvard and the Marin Biological Laboratory at Woods Hole has called "light to hide by."[21] The bobtail squid perform their habitual mating rituals at the ocean surface on moonlit nights, but the down-dwelling light makes their amorous silhouettes visible to predators from below. Microbial luminescence to the rescue! As the vibrios set the light organs of the squid aglitter, their undersides become invisible from below.[22] "*La mer est le ciel des poissons,* " as Jean Cocteau had it.

Vibrios know that there are enough of them to light up the sky by means of quorum-sensing (QS) molecules. These were first identified as members of an *N*-acyl homoserine lactone (AHL) family, but further studies of both gram-positive and gram-negative bacteria found that a vast number of microbes secrete a variety of QS molecules. By now, not only AHLs, but a whole platoon of modified peptides have been identified as signals of mutual aid in microbes.[23,24] As might be expected in our new RNA world, Bonnie Bassler at Princeton has recently shown that a small RNA chaperone protein (Hfq), acting with multiple small regulatory RNAs (sRNAs), functions as "an ultrasensitive regulatory switch that controls the critical transition into and out of quorum sensing mode."[25]

Quorum-sensing molecules launch signaling cascades that cross bacterial species, cut across kingdoms and can serve both benign and pathogenic functions in human disease. Quorum sensing and its extension, the formation of biofilms, have been used to explain human ailments that range from systemic sepsis, to cystic fibrosis, sinusitis, airway infections and even gum disease. Interference with quorum sensing is therefore the next logical step on the way to new antimicrobials.[3,23] Bassler and Losick pointed out that studies of microbial quorum sensing have revealed a new world of "long- and short-range chemical signaling channels; one-way, two-way, and multi-way communication; contact-mediated and contact-inhibited signaling; and the use and spread of misinformation or, more dramatically, even deadly information."[25] They could have been describing signals that did—and mercifully didn't—fill the airways on January 15 when that Airbus splashed down in the Hudson.

A final note. On February 9, 2009, Mayor Michael Bloomberg presented keys to New York City to Captain Sullenberger and his crew, who in turn paid tribute to all those, including Captain Brittany Catanzara, who plucked crew and passengers from the icy Hudson—a story of mutual aid that would have made Kropotkin proud. On April 3, 2009, Bonnie Bassler was awarded the prestigious Wiley Prize at Rockefeller University for her "pioneering investigations into quorum sensing, a mechanism that allows bacteria to 'talk' to one another with chemical languages."[26] Margaret Fuller would have been proud—not only of her sea captain.

# *SiCKO* Statistics:
# Michael Moore and L'École de Paris

*Now, if there is anything on which the biological sciences have prided themselves in these latter years it is the substitution of quantitative for qualitative formulae. If I summed up the lessons of [Professor Pierre C.A.] Louis in two expressions, they would be these:*

*Formez toujours des idées nettes. (Frame your thoughts neatly)*
*Fuyez toujours les à peu pres. (Always avoid approximation)*

—Dr. Oliver Wendell Holmes, 1860[1]

*America's health care system is neither healthy, caring, nor a system.*

—Walter Cronkite, 1993[2]

*There are four times as many health care lobbyists as there are members of Congress. According to the Center for Responsive Politics (www.opensecrets.org), in 2005 there were 2,084 health care lobbyists registered with the federal government. With 535 members of Congress, that's 3.895 lobbyists per member.*

—Michael Moore, *SiCKO*, 2008[3]

## The Numerical Method

IN THE CURRENT DEBATE OVER THE FUTURE of medical care in the United States, we could do worse than to consult two authorities in the field, Professor Pierre C. A. Louis, the father of medical statistics,[4] and Michael Moore, "the angry filmmaker."[5] Moore's *SiCKO* suggests that we have a lot to learn from France; Professor Louis reminds us that it wouldn't be the first time.

Over a century and a half ago, Louis introduced his "numerical method" into the life sciences. He studied 77 patients with febrile pneumonia at La Pitié hospital in Paris and found that the number of those who improved after traditional bloodletting was no greater than of those left alone.[6] He was the first to show that numbers, not intentions, determine whether medical treatment works. Moore's *SiCKO* showed that structure, not money, determines whether a nation's medical system works. Michael Moore tells us that although the United States spent a world-record $5,711 per capita on health care in 2003, only 40 percent of Americans were "satisfied" with its delivery. In contrast, France spent $3,048 per capita and satisfied 65 percent of its population. They also live longer in France: life expectancy in the United States is 77.5, while the French average is 79.6.[3]

Other sources confirm *SiCKO*'s numerical conclusion: the French have us beat by a country mile, *i.e.*, 1.6 km.[7–9] An extensive study by the World Health Organization (WHO) in 2000 ranked France's health care system as best in the world "because of its universal coverage, responsive health care providers, patient and provider freedoms, and the health and longevity of the country's population." The United States clocked in at #37. The United States took a hit due to the high cost of health care, inequality of access, and the growing number of those without insurance—close to 46 million individuals in 2005.[7]

It gets worse. A 2008 study by Ellen Nolte and C. Martin McKee tabulated deaths in 19 countries due to conditions (such as myocardial infarction) in which medical science can make a difference. Again, France ranked first in preventing "amenable deaths" while the United States placed dead last. Nolte and McKee calculated that if the United States performed as well as the top three countries (France, Italy and Australia) there would be over 100,000 fewer "amenable deaths" in the United States each year.[8]

The United States does lead France in one statistic, however. Proportionally, we have more people doing biomedical research. There are 9.6 researchers in the United States for every thousand employees, while France has eight (by comparison, Germany has seven and Mexico one) Does this mean that France is skimping on basic biomedical research in order to boost medical care delivery? Not really. The annual cost of health care in France is 10.5 percent of the country's per capita GDP and French biomedical research gets 0.8 percent. Total American health care costs are 15 percent of GDP per capita and biomedical research is supported at 1.2 percent. Bottom line: both countries spend about the same percentage of their total medical bill on biomedical research, about 8 percent.[9,10] But while research expenditures in the two countries are proportional, the total medical experience is quite different in each.

## A Tale of Two Countries

Sources more rigorous than *SiCKO* confirm that we can learn a lot from France.[11] The present system in France (the Sécurité Sociale) developed after World War II and resulted in complete universal coverage by 2000. This goal was achieved without the nasty public-private wrangle that blocks health care reform in the United States. Nowadays, patients in France are reimbursed an average of two-thirds of all medical costs (hospital, office visits, drugs) directly from Sécurité Sociale. The rest of the cost (set by doctors or hospitals) is reimbursed by one of three large private insurance funds that cover almost all wage earners. These funds collect fees either from employers or individuals, who are charged affordable rates. The private funds, which had their origins in trade and professional insurance co-ops of the last century, are structured like the old Blue Cross/Blue Shield in the United States or the Krankenkassen in Bismarck's Germany. Those who are unemployed or "underemployed" (that is, with income under 6,600 euros a year) are protected by *La couverture maladie universelle*, a government insurance fund that pays their medical costs.[12]

Almost all primary care doctors—70 percent are in private practice—and most specialists have signed up with the Sécurité Sociale. The law guarantees a patient's choice of practitioner and preserves physician autonomy. Unlike the practice in America's insurance systems, no bureaucrat needs to preapprove a procedure, direct a treatment or limit a hospital stay. (I'm reminded of my experiences in the single-payer system of the U.S. Army where the needs of each patient came first without regard to cost, rank or local convenience.) When a French patient sees a doctor of choice, a single microchip-enhanced Sécurité Sociale card permits that doctor immediate online access to the patient's medical chart. That very same card directs reimbursement to the patient's bank account of any out-of-pocket expenses for the visit/consultation/procedure. And unlike Canada or the United Kingdom, there are no long waits for elective surgery.[9] The French are also not required to listen to ads for proprietary drugs during the nightly news. No wonder the majority of patients are satisfied.

Moreover, as Moore shows in *SiCKO*, French doctors remain snug in high-middle-class comfort.[3] While an average American physician makes five times the gross income of his French colleague, the American doctor works under a heavy burden of debt for their premedical and medical education and expenses for their children's home care and day care and for nursery and secondary school fees. In France, each of these is provided by the state. Then, of course, the American doctor is responsible for his own health insurance costs, for prohibitive malpractice insurance and a platoon of clerks required to cope with our byzantine reimbursement schemes. Finally, whereas academic physicians in the United States must raise much of their own income from grants and/or hospital practice, French university professors are employees of the state.

## Learning from Louis

Not only *SiCKO* and the WHO have suggested that we're playing second—or 19th, or 37th—fiddle to France in medical care. A few years ago, *The Boston Globe* published an op-ed praising "France's Model Healthcare System," which concluded that "Perhaps it's time to look at French ideas about health care reform."[9] Bostonians rarely cede preeminence to others in the world of ideas, not to speak of medical practice. Most of them agree with Dr. Oliver Wendell Holmes, who once called Boston's statehouse "the Hub of the Solar System."[13] But let's recall that Holmes himself had gone to Paris to learn French ideas and the numerical method from Professor Louis.

In 1833, Oliver Wendell Holmes arrived in Paris to study the new science, and for a young American the experience was bracing. He wrote to his parents in Boston:

> *I am getting more and more a Frenchman . . . I love to talk French, to eat French, to drink French every now and then (these wines are superb, and nobody gets drunk, except as an experiment in physiology); and I do believe, if Napoleon was alive, and I stayed here much longer, I should want to fight a little.*[14]
>
> *I have become attached to the study of truth by habits formed in severe and sometimes painful circumstance and self-denial. For, trust me, the difficulties in the investigations of our profession . . . the cold and damp and loathsomeness of the dissecting room, are exceedingly repulsive to the beginner.*[15]

But Holmes persevered and, after two years in Paris, returned to Boston with his new microscope, and eventually introduced Louis's numerical method and dissection-based anatomy to the back waters of Back Bay. His temperament squared the Anglophile bent of a Yankee Brahmin (his term) with the new scientific spirit of Paris. He was not the only Bostonian to catch a whiff of France:

> *The more I see of French character, the more I am delighted with it. I have hardly heard an—As I was writing this, [Ralph] Waldo Emerson came up to see me. He had been sitting some time when I heard another knock, and in walked—James Russell [Lowell]! I never was so astonished in my life; but as he is here, and I must attend to him, without ceremony I shall take the liberty to conclude my letter . . . Give my love to everybody.*[16]

The literary Flowering of New England followed upon the commercial success of Boston after the American Revolution, and similarly, the scientific rise of the École de Paris followed the intellectual ferment of Paris after the French Revolution and Napoleonic era. If the large central hospitals of Paris, which brought Holmes to Paris, were a legacy of the Revolution and of Napoleon, the new scientific spirit was a legacy of a young genius named Xavier Bichat (1771–1802). Bichat's theory that disease resides not in organs but in tissues, conquered Europe and was brought to American shores by doctors from Boston, Philadelphia and New York.[17] Years later, Holmes asked:

> *And is it to be looked at as a mere accidental coincidence, that while Napoleon was modernizing the political world, Bichat was revolutionizing the science of life and the art that is based upon it; that while the young general was scaling the Alps, the young surgeon was climbing the steeper summits of unexplored nature; that the same year read the announcement of those admirable "Researches on Life and Death," and the bulletins of the battle of Marengo?*[18]

Holmes looked back at that time of change in Paris and concluded that there was the "closer relation between the Medical Sciences and the conditions of Society and the general thought of the time, than would at first be suspected."[19] In Paris, after the Napoleonic decades and the 1830 revolution, reform was in the air—medical no less than social. Following the example of Bichat, the École de Paris launched a medical revolution that would elevate study of the human body and its ailments to an exact science, to explore "the steeper summits of unexplored nature." Their goal was a system of human biology as grand and coherent as Lamarck's evolutionary biology or Cuvier's animal kingdom at large.

Before the 19th century, it had been uncommon for doctors to record their findings quantitatively. But, again following Bichat, clinicians began to assign numbers to a patient's vital signs—the pulse, temperature and respirations—and then in 1819 came René Laennec's stethoscope. After pulse and temperature were routinely noted, after heart sounds were recorded and analyzed, doctors could begin to settle their puzzling cases not by conjecture but by comparing them to similar cases with similar numbers. By this means, Louis finally ended therapeutic blood-letting in modern hospitals. Holmes was right on when he wrote of his mentor's work, "I consider his modest and brief Essay on Bleeding in some Inflammatory Diseases, based on cases carefully observed and numerically analyzed, one of the most important written contributions to practical medicine, to the treatment of internal disease of this century . . . it was a revolution."[20]

The spirit of the École de Médecine revolution is captured by Holmes in a letter that explains to his friend John Osborne Sargent why he cannot write poems for *The New England Magazine*:

> . . . it is always with my note book in my hand; that I often devote nearly two hours to investigating a difficult case, in order that no element can escape me, and that I have always a hundred patients under my eye. Add to this the details and laborious examination of all the organs of the body in such cases as are fatal—the demands of a Society [Louis's Society of Clinical Observation] of which I am a member—which in the course of two months has called on me for memoirs to the extent of thirty thick-set pages—all French, and almost all facts hewn out one by one from the quarry—and No, John, a heavier burden from my own science, if you will, but not another hair from the locks of Poesy, or it will be indeed an ass's back that is broken.[21]

**Universal Coverage and the Morgue**

In Paris, Holmes studied the diseased organs of "such cases as are fatal" in dissecting rooms where victims of murder, suicide or accident joined those felled by disease. The bodies were gathered from the dead-house on the rue Morgue to become the stuff of medical knowledge at the École de

Médecine further down the Seine. A chilling etching by Charles Meryon shows the quayside mortuary ready to receive a night's cargo from scows on the river; unwanted bodies were sold for cash to students of medicine eager to practice surgery, learn anatomy or venture into forensics. Our word, "morgue" comes from that dead-house by the Seine.

On the other side of the Atlantic, students had few opportunities to dissect the human body. Holmes explained that, after tutelage in Paris, "the attentive student may return a sounder physician at twenty-five than many who slumber till sixty in our own languid scientific atmosphere."[22] But parochial piety held sway in America to midcentury: in the 1820s the Vermont Academy of Medicine proclaimed that "bodies disinterred hereabouts would not be used in the department of practical anatomy," and as late as 1845, only half of American medical schools required time in the anatomy lab.[23] The difference between the American experience and the French is spelled out by 24-year-old Oliver Wendell Holmes:

> *The whole walls around the École de Médecine are covered with notices of lectures, the greater part of them gratuitous; the dissecting rooms, which accommodate six hundred students, are open; the lessons are ringing aloud through all the great hospitals. . . . It is an odd thing for anybody but a medical student to think of, that human flesh should be sold like beef or mutton. But at twelve o'clock every day, the hour of distribution of subjects, you might have seen Bizot and myself—like the old gentlemen one sometimes sees at a market—choosing our days' provision with the same epicurean nicety. We paid fifty sous apiece for our subject, and before evening we had cut him into inch pieces. Now all this can hardly be done anywhere in the world but at Paris. . . . I have told you all this to let you know that I am not staying at Paris for nothing.[24]*

Michael Moore's film and his subsequent web postings have made many Americans aware that our health care delivery system in 2009 is as far behind the rest of the world—and especially the French—as American medical science was behind its French counterpart in 1830. The reasons are not dissimilar: parochial piety and our provincial belief that nowhere else are things better done than here.

By the time Holmes had settled into French ways, he asked his parents in Paris,

*How could I ever have dined at two o'clock? How could I have put anything to my mouth but a silver fork? How could I have survived dinner without a napkin? . . . It is very narrow and ridiculous, and yet it is very common, to hear people taking the standard of their own fancy for that of necessity. [An American] will tell you that he prefers a separate plate from his neighbor, but has no idea of any napkin but the tablecloth— another would shudder at an iron tumbler, but is astonished that his neighbor has an aversion to an iron fork. Now as for napkins and silver forks, the most ordinary, meanest eating houses in Paris consider them as indispensable,—and so with regard to many things which we consider as luxuries, they make a part of ordinary existence with the Parisian.*[25]

More and more Americans are convinced that our broken health care system is as outmoded as the use of iron forks or a tablecloth used as an alternative for a napkin or two. More and more Americans also agree with Michael Moore and *SiCKO* that universal health care coverage should be as much a part of ordinary existence for Americans as it has been for Parisians for over a decade.

# Ask Your Doctor:
# Justice Holmes and the
# Marketplace of Ideas

*When men have realized that time has upset many fighting faiths, they may come to believe . . . that the ultimate good desired is better reached by free trade in ideas—that the best test of truth is the power of the thought to get itself accepted in the competition of the market.*

—Oliver Wendell Holmes Jr., 1919[1]

*The liberty of the citizen to do as he likes so long as he does not interfere with the liberty of others to do the same . . . is interfered with by school laws, by the Post Office, by every state or municipal institution which takes his money for purposes thought desirable, whether he likes it or not. The Fourteenth Amendment does not enact Mr. Herbert Spencer's* Social Statics [*of* laissez faire].

—Oliver Wendell Holmes Jr., 1905[2]

*This survival of the fittest, which I have here sought to express in mechanical terms, is that which Mr. Darwin has called "natural selection," or the preservation of favoured races in the struggle for life.*

—Herbert Spencer, 1864[3]

### Survival of the Fittest on the Tube

THESE DAYS, JUSTICE HOLMES would be quite at home watching today's marketplace of ideas: the evening news on network television. He'd surely appreciate that every commercial for a prescription drug is a compromise between free-market laissez-faire and government regulation. Thanks to U.S. Food and Drug Administration (FDA) rules framed in 1997, those direct-to-consumer advertisements don't only promise us a leg-up in the struggle for existence. They are also required to warn us of any possible side effects and above all, to "Ask your doctor." That's how we've come to dread those "rare but serious" muscle cramps from statins, that "rare but serious condition called TTP" from an antiplatelet agent and the blockbuster erection that "lasts longer than four hours."

Only in the United States and New Zealand are manufacturers of what used to be called ethical drugs permitted to advertise directly to consumers. That example of American exceptionalism caught the attention of a French visitor to our shores recently. After watching American network news between six and seven p.m., he concluded that our nation no longer manufactures wristwatches or washing machines but makes money by selling drugs to folks with urinary incontinence, osteoporosis, clogged arteries and erectile dysfunction. He wondered why Americans are instructed to "tell your doctor if they have liver disease, hypertension, or glaucoma—or if they are pregnant." In the rest of the world, that's what doctors tell patients. He wondered why a patient was told to "call your doctor if you experience a sudden loss of vision" after taking an erectile dysfunction pill: "If you suddenly can't see," he asked, "how do you know where you left the cell phone?"

It's a different story in France. The television commercials during its nightly news (between seven and eight p.m.) persuade one that the nation makes lots of watches and washing machines and sprays perfume over every lovely poitrine in the land. No doctor is summoned: the French are told to ask their *chocolatier* for liqueur-filled bonbons, their *fromager* for running cheese and *demandez à votre boucher* some ham from Bayonne. No side effects either: the French are thinner—a body mass index of 23.34 in France versus 28.22 in the United States,[4] their life expectancy is longer—80.9 versus 78.1 years[5] and they spend less money annually on drugs—$599 versus $752 per capita.[6] Holmes and Spencer would have had little trouble deciding which country was favored in the struggle for life.

### A Drug on the Market

The jury is still out on whether all of those advertisements are useful for patients, doctors or the pharmaceutical industry. Numbers first: it's estimated that the United States will spend approximately $290 billion for prescription drugs in 2009.[7] To earn that amount, the industry spends between $40 and $60 billion on promotion, depending on how and by whom it's calculated.[8,9] Critics point out that over the last few years, the promotional costs of prescription drugs in the United States have consistently been twice those of research and development, but that's another matter.[9]

In the decade after the FDA opened the airways to ask-your-doctor advertisements on television (1997), spending on these increased over 300 percent; was expected to reach over $4 billion in 2009.[10] But, that is still only 14 percent of overall expenditures on drug promotion, so it is not a big deal compared with drug reps' free samples, professional education, etc.[11] Finally, not all drugs are peddled equally on the tube; it's the survival of the fittest, after all. In 2005, for example, manufacturers of proton-pump inhibitors, statins and erythropoietins spent fully one-third of their promotional budget on direct-to-consumer ads, some as much as Budweiser or Nike in one year.[11] We don't yet know who will survive as fittest in the erectile dysfunction game.

How is that game played? By hitting those at risk. A quantitative study from Emory University documented that:

> *Direct-to-consumer drug ads appeared most frequently during news programs and soap operas and during the middle-afternoon and early-evening hours [that] may be targeted specifically at women and older viewers.*[12]

The authors concluded that the average television viewer is exposed to more than 30 hours of direct-to-consumer pharma advertisements per year, far surpassing any exposure to other forms of health communication.[12] And, sure enough, the results are clear: the ads have been found to make a positive difference in sales, in requests for specific agents and in the quality of encounters between doctors and patients.[8,11,13] Indeed, about half of primary care physicians and an equal number of patients report that a new prescription had been influenced by a television advertisement.[8,11] It is no surprise then that by 2006, the American Medical Association and the American College of Physicians had come out for a complete moratorium or for tighter regulation of the ask-your-doctor advertisements.[8]

**Regulation Means Prohibition**

The impetus to have companies advertise directly to consumers, rather than leaving the choice of drugs to the doctor, started in the 1980s. The libertarian left argued for "patient empowerment," and the entrepreneurial right welcomed the freedom of market competition.[14] However, if those motives were appropriate then, are they still appropriate today? We've surely learned what a lack of regulation can do to mortgages, financial institutions and the "health care" industry. Haven't we gone overboard with those ask-your-doctor ads, with laissez-faire in the erectile dysfunction market? American marketing techniques may well have subverted the doctor-patient relationship more than in countries with universal health insurance and a public option where folks ask their butcher for ham and a doctor tells them if they are pregnant. Those advertising dollars cost all of us money: the 40-odd billion dollars per year for drug promotion seems a tad high these days. Couldn't we save money by cutting out all of that unnecessary advertising by pharmaceutical companies, hospitals and individual doctors? Why does a teaching hospital in Pittsburgh advertise in *The New York Times*, which serves a city not entirely devoid of academic centers? France seems to do pretty well without full-page advertisements in *Nice-Matin* for the Hôpital Saint-Antoine in Paris.

We turn to Justice Holmes, who was there at the beginning of drug regulation in this country. The original Food and Drug Act of 1906, passed in response to patent medicines that were laced with cocaine,[15] simply required pharmaceuticals to list each ingredient, fully and honestly, on their labels. And any medicine, be it for croup or cancer, could make any claim it wished for clinical efficacy. Upholding the act, Justice Holmes ruled for the majority of the Supreme Court (*United States v. Johnson*) that a statement on the label of a bottle of medicine that the contents are effective as a cure for cancer, even if misleading, is not covered by the statute.

> *[Congress] was much more likely to regulate commerce in food and drugs with reference to plain matter of fact, so that food and drugs should be what they professed to be, when the kind was stated, than to distort the uses of its constitutional power to establishing criteria in regions where opinions are far apart.*[16]

Conclusion: The first FDA act guaranteed only the content, not the efficacy, of a medication. Ingredients were facts; efficacy fell into the "regions where opinions are far apart" that were best contended in the marketplace. The justice was probably influenced in this opinion by his father, Dr. Oliver Wendell Holmes, who (not incorrectly) cautioned the Massachusetts Medical Society in 1860:

*Throw out opium . . . throw out wine, which is a food, and
the vapors which produce the miracle of anesthæsia, and I
firmly believe that if the whole materia medica, as now used,
could be sunk to the bottom, and all the worse for the fishes.*[17]

Half a century passed before regulatory control was imposed. Insulin,
vitamin B$_{12}$, sulfa drugs and other effective medicines moved into the clinic,
but the *safety* of a product did not become requisite for FDA approval until
1938. Efficacy of a drug for a clinical indication was not required as a crite-
rion until 1962.[18] It's been pointed out in these pages that unfortunately,
Holmes's ruling of 1906 still applies with respect to such "dietary supplements" as echinacea, gingko biloba and St. John's wort. All that's required
to peddle these on the air or the web is a label on the bottle stating their
ingredients: claims for the efficacy are left to regions of opinion.[15]

For most of his career, Justice Holmes was torn between the social Darwinism of the Spencer variety, with its emphasis on the survival of the fittest, and the very Bostonian notion of the City on a Hill. He formulated the
principle of the marketplace of ideas, actually the *"free trade in ideas,"* in his
dissent re: *Abrams v. United States.*[19] The majority of the court upheld the
wartime Espionage Act of 1917, finding guilty a left-wing activist who had
flung revolutionary leaflets in English and Yiddish from tenement windows.
Holmes dissented:

*I think that we should be eternally vigilant against attempts
to check the expression of opinions that we loathe and believe
to be fraught with death, unless they so imminently threaten
immediate interference with the lawful and pressing purposes
of the law that an immediate check is required to save the
country.*[1]

Holmes sounded this tocsin of libertarian values in a scoundrel time
of jingoism, as U.S. forces were sent overseas on the side of anti-Bolshevik
forces in postwar Russia. (The court's majority opinion about the Espionage
Act remained in force until the Vietnam War.)

However, Holmes *could* support a reasonable degree of regulation;
federal law could be used to check Mr. Spencer's law of laissez-faire. In
1918, the Supreme Court majority ruled that free trade would be unduly
restrained if an upstate New York baker couldn't order his underage employees to work a ten-hour shift (*Hammer v. Dagenhart*). In a dissent that laid
the groundwork for child labor laws, Holmes wrote:

*Regulation means the prohibition of something, and when in-*
*terstate [child labor] commerce is the matter to be regulated,*
*I cannot doubt that the regulation may prohibit any part of*
*such commerce that Congress sees fit to forbid. The public pol-*
*icy of the United States is shaped with a view to the benefit of*
*the nation as a whole.* [20]

## Three Generations of Imbeciles

Indeed, despite the marketplace analogy, which has so often been associated with his writings, I'm unsure whether Holmes would join the
American Medical Association and the American College of Physicians in
a drive to limit those pesky ask-your-doctor advertisements. On the one
hand, he admired Spencer's principle of the free market; his Civil War
combat experience with the 20th Massachusetts (Harvard) Regiment persuaded him that the fittest survive in fierce competition. He had written
to Lady Pollock in 1895:

*You English never quite do justice to Herbert Spencer. And*
*yet after all abatements I doubt if any writer of English except*
*Darwin has done as much to affect our whole way of thinking*
*about the Universe.* [21]

Social Darwinism, à la Spencer, was celebrated at the centenary of the
birth of Charles Darwin in 1909. In a gala ceremony, the University of
Cambridge awarded honorary degrees to disciples of Darwinism on both
sides of the Atlantic. German biology was represented by Ernst Haeckel
and August Weismann, and Anglo-Saxon jurisprudence was honored when
Justice Oliver Wendell Holmes of the United States was awarded the honorary degree of Doctor of Civil Laws. [22] The new century began on a note of
triumph for the preservation of favored races in the struggle for life.

In accord with that principle of favored races, Holmes might weigh in
on the side of regulating those ask-your-doctor advertisements. A product
of three generations of intellectual entitlement, he might question the wisdom of leaving the choice of medicines to a viewer of daytime television
rather than a graduate of the Harvard Medical School. Holmes and his
generation appreciated medicine as a learned profession, along with theology and law. And three generation of the Holmes family had been leaders
in these professions: Reverend Abiel Holmes of Cambridge (1763–1837),
Dean Oliver Wendell Holmes of Harvard Medical School (1809–1894)

and Holmes himself. Likewise, Holmes was sure that genetic merit must have guaranteed the success of three generations of the Darwin family, each a Fellow of the Royal Society: Erasmus Darwin, MD (1731–1802), the natural philosopher; Robert Darwin, MD (1766–1848), a king's physician and Charles—well, Charles Darwin.

In 1927, Justice Holmes wrote the majority opinion in the case of *Buck v. Bell*, arguing for the strictest form of government regulation: that of heredity. Carrie Buck was a "feeble minded white woman" who was committed to the State Colony for the mentally infirm in the Commonwealth of Virginia. She was the daughter of a "feeble minded" mother in the same institution and the underaged mother of yet another "illegitimate feeble minded child." An Act of Virginia, approved March 20, 1924, ruled that "the welfare of society may be promoted in certain cases by the sterilization of mental defectives, in males by vasectomy and in females by salpingectomy."[23] Holmes was convinced that "heredity plays an important part in the transmission of insanity, imbecility, etc." Recalling his own wounds and the Harvard dead at Antietam, the justice ruled:

> *We have seen more than once that the public welfare may call upon the best citizens for their lives. It would be strange if it could not call upon those who already sap the strength of the State for these lesser sacrifices. The principle that sustains compulsory vaccination is broad enough to cover cutting the Fallopian tubes. . . . Three generations of imbeciles are enough.*[23]

Compared with that distinction between the "best citizens" and "those who sap the strength of the state," the struggle between laissez-faire and regulation in the ask-your-doctor debate seems like a petty argument indeed. I'd bet that Justice Holmes would opt for the free competition of the market—with the public option of regulation.

# Filter the Dogs:
# Microbial Mishaps in Massachusetts

*Drinking Water Warning: On June 15, 2010 at approximately 3:00 pm, the Mass DEP issued a boil order for the Town of Falmouth. . . . The two areas of primary concern are the Woods Hole and North Falmouth Fire Stations. We took 44 samples for coliform bacteria during June 2010. 16 of those samples showed the presence of coliform bacteria. In addition, 5 of those samples showed the presence of* E. coli *bacteria.*

—Town of Falmouth, MA, June 16, 2010[1]

*They didn't have money for a filtration system for the town's water supply? They let dogs loose around Long Pond? Filter the dogs!*

—Eugene P. Kennedy, June 18, 2010[2]

Escherichia coli *strain K-12 is arguably the single organism about which the most is known. [It is] the primary model organism for basic biology, molecular genetics and physiology of bacteria, and was the founding workhorse of the biotechnology industry.*

—Monica Riley, Marine Biological Laboratory,
Woods Hole, MA, 2006[3]

*The bills of mortality are more obviously affected by drainage than by this or that method of medical practice.*

—Oliver Wendell Holmes, 1863[4]

## The Macro- and Microbiome

THE BILLS OF MORTALITY from microbial disease have been affected not only by drainage, by the sanitarians, but also by bench scientists, the microbe hunters. The summer of 2010 witnessed a unique sanitary mishap in the very home of microbe hunters. After coliform bacilli and *Escherichia coli* were found at several sites in the water supply, a "boil order" was issued by the town of Falmouth, Massachusetts, which supplies the village of Woods Hole, Massachusetts, with its drinking water.[5] Water fountains went dry at the Marine Biological Laboratory (MBL), site of a major microbial sequencing effort to define all of the genes and proteins of *E. coli*.[3]

Most of Falmouth's water supply comes straight from Long Pond, a rustic, mile-and-a-half-long body of water that abuts the center of town. A "leg-stretching" hiking trail, rich in wildlife, winds around the edge of the pond. It welcomes bikers, campers and lots of folks walking lots of dogs.[6] The water from Long Pond remains unfiltered on its way to the tap: "All we do is smack it with chlorine. . . . We are basically drinking pond water which is very susceptible and vulnerable to bacteria," explained Falmouth's water superintendent.[7]

In 1993, the town had received a waiver from the state that permitted Falmouth's selectmen to smack the pond with chlorine rather than to raise the $20 million or so in taxes for a filtration system required by federal law. Each summer, therefore, when Falmouth's population triples from its winter level of 32,000, Long Pond is routinely awash in "algae, microorganisms, inorganics and decaying organic matter."[8] Add seasonal thunderstorms, and the pond's ecosystem—its *macrobiome*—is transformed.[9]

On the weekend of June 5, 2010, heavy rainstorms diluted the drinking water's chlorine levels to below antimicrobial strength; rainwater also leached fecal debris of man and beast into the unfiltered water system. "Filter the dogs!" quipped Harvard's Eugene Kennedy.[2] *E. coli* and coliform counts in excess of "normal" were detected on June 7, but the boilwater order was not issued until June 15. Result: gram-negative rods and who knows what else flowed freely from taps in the town. Fans of Lake Wobegon will understand the reasons for delays of this sort: local politics, importune vacation schedules and simple ignorance of the rules. At any rate, after a week of sewage in the taps, the order proved effective. Water was boiled, the pond was doused in chlorine and sales of bottled water soared. The water superintendent and his deputy were suspended.[7,8] A. Sidney Knowles, a scholar from Raleigh, North Carolina, suggested that the two officials could have raised money for a filtration plant by peddling pond water as PEPS*E. COLI*.[10]

Coliform bacteria are part and parcel of all of us; they constitute much of our *microbiome*, first defined by Joshua Lederberg: "Just as scientists study entire ecological systems to see how the various parts interact, we must [study] what I call the microbiome: the menagerie of the body's attendant microbes."[11] Such microbes are rarely pathogenic: *E. coli*, for example, don't survive for long in water. However, as the MBL's Mitchell Sogin explained in *The Falmouth Enterprise*, coliform organisms are useful "indicator organisms" to warn water inspectors that other fecal pathogens—from *Shigella*, *Salmonella* and *Campylobacter* to *Giardia* and *Entamoeba* —may have reached the water supply.[12] Citing new work about DNA sequencing of water samples, Sogin predicted that greatly improved genomic methods to detect real pathogens should soon ensure safe water in our ponds, streams and oceans. He pointed out that "The U.S. government, for instance, is now funding four Centers for Oceans and Human Health, one of which is located in Woods Hole and represents a collaboration between the MBL, Woods Hole Oceanographic Institution and MIT."[12] It's a noble effort, but it's unlikely that any level of advanced microbe hunting can compensate for local mishaps as a result of fumbling, ignorance or penny-pinching.

### Coli at the Seaside

The real paradox of the Woods Hole boil-water order is that *E. coli* has long been the subject of study by the most modern of microbe hunters, many of whom have worked at the MBL. Selman Waksman, who discovered streptomycin and coined the term "antibiotic," began his work at Woods Hole in 1931; his gravesite is in the village.[13] In the summer of 1947, Joshua Lederberg wrote his PhD thesis at the MBL library: his paper with E. L. Tatum, "Gene Recombination in the Bacterium *Escherichia coli*," was part of their Nobel citation in 1958. It was the dawn of microbial genetics.[14] In 1958, at Woods Hole, Matthew Meselson and Franklin Stahl framed "the most beautiful experiment in biology": the DNA of *E. coli* replicates in a semi-conservative manner.[15] (So does our own.) In the 1950s and '60s, Arthur Pardee and Jean-Pierre Changeux, long-time MBL visitors, founded much of molecular biology on the back, so to speak, of *E. coli*.[16,17] In 1966, Sir Hans Kornberg described *anaplerosis*, a mechanism *E. coli* use to recharge their metabolic battery.[18] In the 1980s,

Eugene P. Kennedy figured out why *E. coli* has trouble in the pond. Kennedy recalls: "Driving home to Brookline from Woods Hole one evening, I [asked] how does a cell detect the osmolarity of the medium in which it is growing?[19] His answer: membrane sugars and lipids permit *E. coli* to handle the osmotic trauma of leaving our gut to end up in the can.[20] However, they can't recharge their batteries in the pond, and that's why they're just indicator organisms.

To round things out, in the summer of 2000, Lederberg launched a new program for the Ellison Medical Foundation; at its annual MBL retreat, he defined why it was important to explore *E. coli* and its fellow inhabitants of a new domain. The human body, he argued, should be considered "a superorganism with an extended genome that includes not only its own cells but also the fluctuating microbial genome set of bacteria and viruses that share that body space," its microbiome.[21]

And sure enough, Monica Riley, with other MBL scientists in Mitch Sogin's Bay Paul Center, are working out the genetic encyclopedias of *E. coli* and other members of our microbiome,[3] an effort that would have made Josh Lederberg proud. His concept of the microbiome has now been widely appreciated, from Woods Hole to the National Institutes of Health Human Microbiome Project.[22]

### The Little Army of Microbes

Glory for the microbe hunters, mishaps for the sanitarians. All that science, all those brains, all those millions spent on labs, machines and IT networks! Meanwhile, coliforms invade the drinking water because a few selectmen didn't raise taxes to pay for proper sanitation or "drainage," as Dr. Oliver Wendell Holmes Sr. had it. His son, Justice Oliver Wendell Holmes Jr., got it right: "Taxes are what we pay for a civilized society"[23]—or a clean one, he might have added.

Dr. Holmes set the standard for sanitary reform in Massachusetts. Years before Pasteur demonstrated that puerperal sepsis was caused by the streptococcus, Dr. Holmes had worked out how childbed fever was spread and how it could be prevented. Late in life, he wrote to William Osler:

> *I do know that others had cried out with all their might against the terrible evil [of puerperal sepsis] before I did and I gave them full credit for it. But I think I shrieked my warning louder and longer than any of them and I am pleased to remember that I took my ground on the existing*

*evidence before the little army of microbes was marched up
to support my position.*[24]

Holmes's cry was based on another microbial mishap in Massachusetts. Young Dr. Holmes had just returned to Boston from medical studies
in Paris. He had trained at the large, public hospitals, where the "birth
of the clinic" anteceded the sanitary revolution. In those Paris hospitals,
he had observed doctors in street clothes examining women in labor, de-
livering their babies and rendering puerperal care from bed to bed. They
often moved on to the next woman without washing their hands. In Boston, where there were few public beds, babies were home-delivered, but
similar practices obtained. A local practitioner wrote to Holmes about a
"disastrous period in my practice," when between July 1 and August 13,
1835, he attended at the delivery of 14 women; eight of these developed
puerperal fever, and two died. He admitted that he had not changed his
outer coat until August 6, the date of confinement of the last victim![25]
Holmes studied two other such focal outbreaks and in 1843, presented
his conclusion in "The Contagiousness of Puerperal Fever" to the Boston
Society for Medical Improvement. He argued that:

*The disease known as puerperal fever is so far contagious as
to be frequently carried from patient to patient by physicians
and nurses.*[25]

Holmes reviewed the British Registrar's statistics. These estimated
that the mortality in England from puerperal sepsis in the 1830s varied
from three to five out of 1,000. He contrasted this low, sporadic level
with American outbreaks arising from one focus: Holmes commented,
"in view of the general incidence of the disease being less than five in a
thousand, it does appear a singular coincidence, that one man should have
ten, twenty, thirty or seventy cases of this rare disease following his or her
footsteps with the keenness of a beagle."[25]

The countermeasures Holmes urged were those of the sanitarian revolution: there should be plenty of soap and water and destruction of contaminated clothing, linens and blankets, and doctors or nurses ought not
to deliver babies for several months after an outbreak. Yet, Holmes's sanitary warning fell on deaf ears. Who was this Holmes after all? He was a
young doctor, just out of training, and from Massachusetts, then a remote
corner of the medical world. In Europe, Semmelweis, Lister and Pasteur
went on their microbial way without attention to a Bay State backwater.

General inattention to his earlier warning led Holmes to republish and elaborate his work about puerperal fever in 1855.[26] By then, Holmes had become a stronger force in the profession; he was not only the Parkman Professor of Anatomy at Harvard Medical School but also dean of its medical faculty. (It was Holmes who, in 1846, had given the name of "anesthesia" to Morton's discovery.) He had also become a respected poet and lyceum lecturer; his monthly essays in *The Atlantic Monthly*, collected in *The Autocrat of the Breakfast-Table,* had already made him a household name among cultured folks on both sides of that ocean. In his new pamphlet, he directed harsher words at his main opponents and marshaled statistics against them:

> *I have had the chances calculated by a competent person, that a given practitioner, A., shall have sixteen fatal cases in a month, [data from Philadelphia records] it follows there would not be one chance in a million million million millions that one such series would be noted.*[26]

Holmes's peroration was memorable:

> *I am too much in earnest for either humility or vanity, but I do entreat those who hold the keys of life and death to listen to me also for this once. I ask no personal favor; but I beg to be heard in behalf of the women whose lives are at stake, until some stronger voice shall plead for them.*[26]

This time, the case was overwhelming, as was its acceptance by the American medical community. Dr. Fordyce Barker commented that in his public hospital, Bellevue, there were times when "a pulse of 120 and a flushed cheek were looked for as a matter of course on the morning after confinement, and the normal results were luxuries to the attendant physician."[27] However, in 1857, in the same hospital, Holmes's strictures were applied after an outbreak of childbed fever. Infected women were removed, the place cleaned, new furniture installed and a different set of doctors and nurses placed in charge. The following month, 23 women were delivered, not one of whom contracted the fever. Dr. Barker not only agreed with Holmes that the physician was often the medium of infection but that "the work of Ignaz Semmelweis in Vienna fully bore out Dr. Holmes's observations."[27]

Sanitary reform had planted its flag in America, and public hospitals soon followed. Boston City Hospital was established in 1864; it was among the first to lay down Holmes's hand-washing practice as law in the delivery room. Score one for the East Coast sanitarians, a decade and a half before Pasteur chalked an image of the "little army of microbes" on a blackboard in Paris.[28]

What happens at childbirth in home and hospital goes hand in hand with what happens in the water supply. These days, only three of every 100,000 women in the industrialized world die in the puerperum, a rate that would have astounded the sanitarians and microbe hunters of yesteryear. Waterborne diseases are also at an all-time low, thanks to the sanitarians. However, in countries that lack sanitation, antibiotics or mandatory filtration systems for their water (Malawi, Chad, Somalia), the rates of maternal mortality and waterborne outbreaks of disease are more than 100-fold higher.[29,30]

Taxes are not the only price we pay for a civilized society, but they could surely help pay for clean water in Massachusetts, where for a week, the townfolks were as much at risk as if they lived in Somalia.

# Pattern Recognition and Gestalt Psychology: The Day Nüsslein-Volhard Shouted *"Toll!"*

*Probably I just shouted: "Toll!" in a conversation with Eric Wieschaus, when he and I scored the fixed embryos together on a discussion microscope, when we shared a lab in Heidelberg. It was during the mutant screen in which we discovered the segmentation mutants in late 1979. This work was published in 1980 in* Nature, *and is the essential basis for our Nobel award.*

—Christiane Nüsslein-Volhard, Personal communication
to the author in *The FASEB Journal*, May 14, 2010

| | |
|---|---|
| Interviewer: | Hat es in Ihrem Leben den Heureka-Moment gegeben? *(Have you ever had a Eureka! moment in your life?)* |
| Nüsslein-Volhard: | Immer wieder mal. Das ist ganz toll! *(Again and again. That's what's so crazy!)* |

—Christiane Nüsslein-Volhard, 2009[1]

*The fundamental of the Gestalt "formula" might be expressed in this way: There are wholes, the behavior of which is not determined by that of their individual elements, but where the parts are themselves determined by the innate nature of the whole.*

—Max Wertheimer, Address to the Kant Society, 1924[2]

*[Mountcastle, Hubel and Wiesel] confirmed the inferences of the Gestalt psychologists by showing us that . . . the brain does not simply take*

*the raw data it receives through the senses and reproduces it faithfully. Instead each sensory system first analyzes and deconstructs, then restructures the raw incoming information according to its own connections and rules . . . shades of Immanuel Kant!*

—Eric Kandel, *In Search of Memory*, 2006[3]

## The Shape of Things to Come

LATELY, MUCH OF HUMAN PATHOBIOLOGY has been chalked up to Toll-like receptors (TLRs), molecules that tell our innate immune system that there is a stranger at the door. In 2010, one medical journal alone has published papers on the role of TLRs in stem-cell activation, dengue fever, bacterial sepsis, microRNA processing, release of leukotrienes and—no kidding—obesity. Other recent reports have implicated TLRs in human ailments that range from gingivitis to Lyme disease; from lupus to titanium hips, and from atherosclerosis to respiratory ailments brought on by the World Trade Center disaster.[4–7]

Toll-like receptors have now been identified up and down the great chain of being: plants, bacteria, sponges, worms, flies and us.[8,9] In humans, TLRs and other "pattern recognition receptors" warn us against generic bacteria, viruses, protozoa or even splinters in our fingers. But they don't tell us whether we've met an individual microbe before: they just assist the work of antibodies and cells of acquired immunity.[4,10] As an abbreviation, TLR is fast becoming as famous as RNA or DNA: PubMed lists over 17,000 papers devoted to TLRs.

But where did "Toll" come from? It's been properly credited to Christiane Nüsslein-Volhard, and first appeared in print as the "Toll gene product" in two of her 1985 articles in *Cell*.[11,12] However, no firsthand account had been rendered of how, why or when this peculiar name was chosen. The word "Toll" appears in neither her Nobel Prize–winning paper from Heidelberg with Eric Wieschaus[13] nor her eloquent Nobel lecture.[14] To complete the record, I wrote to Dr. Nüsslein-Volhard; in reply, she owned that "the original discovery has nothing to do with its present fame," (personal communication from C.N-V May 14, 2010). She is too modest: a gene and gene product that tells an organism which way is up or down has everything to do with Toll's present fame. It's the Gestalt that counts.

On the day that Christiane Nüsslein-Volhard shouted *"Toll,"* she and Eric Wieschaus were working out the genetic control of morphogenesis in the fruit fly, *Drosophila melanogaster*. Insects are segmented, and by studying the patterns of segmentation in many mutants of fly eggs, the Heidelberg scientists were able to identify 15 gene loci that control its final form, its Gestalt. These genes act at three levels of spatial organization. The first set of genes dictates the pattern of segmentation of the entire egg along the body axis: anterior/posterior and dorsal/ventral. The next group of genes governs development of every second segment, while the third group of genes fine-tunes the structure of each segment in the adult.[14,15] The Toll gene tells the developing larva which is front and which is back, and the Toll mutant gets it wrong.

Dr. Nüsslein-Volhard wrote us that:

> *The Toll mutant was so fascinating because we knew already the dorsalised phenotype, and I just had described "dorsal" which displays this polarity loss along the dv axis in lack of function alleles. Toll was the first mutant with a ventralized phenotype, and initially we were tempted to call it "ventral." But the first name got stuck.*

The mutant was all back and no front: Toll, indeed! But words shouted in the heat of discovery have more than their dictionary meaning. My émigré father used "toll" when he meant "crazy," but also "curious" or "amazing"; he used it when he first treated a patient with cortisone. These days German speakers also use toll instead of "cool" or "droll," "outrageous" or "awesome." So toll has all sorts of meanings, but none more profound than when first applied to the "ventralized phenotype:" the topsy-turvy fruit fly in Heidelberg.

Her letter continues:

> *We had bets if it would fit into the dorsalisation series pathway, (which it did, according to the subsequent genetic analysis performed mainly by Kathryn Anderson and Gerd Jürgens as postdocs in my laboratory at the FML in Tübingen). The polarity loss of the original Toll allele is not complete, and Toll has recessive alleles (null alleles recovered as revertants) which show a completely dorsalised phenotype.*[11]

After Kathryn Anderson had gone on to clone the Toll gene, it looked quite a bit like the human interleukin 1-receptor, (IL-1R).[16,17] Since IL-1 is an endogenous pyrogen—it causes fever in man and beast—the next question was how a signal for spatial orientation in flies could possibly explain features of innate immunity: that's what a good number of those 17,000 papers are all about.[18] They suggest that the Heidelberg lab was on the right track at the very beginning. When you turn a creature topsy-turvy, you've changed its overall pattern—its Gestalt.

## Gestalt Patterns

Wieschaus explained that "Christiane and I were both primarily interested in spatial patterns." Sure enough, in his Nobel lecture "From Molecular Patterns to Morphogenesis: The Lessons from Drosophila," the words "pattern" and "patterns" appear 43 times.[15] But Nüsslein-Volhard beats him out in the pattern-count: in her Nobel lecture the words appear 57 times![14] Their Nobel Prize might be said to recognize the importance of pattern recognition.

As Nüsslein-Volhard wrote, TLRs are certainly more "famous" today in immunology than in developmental biology.[11] It turns out that the genes that determine the Gestalt of fruit flies also dictate innate immunity in mice and men. The intracellular signaling pathway is also ubiquitous, it begins with MyD88 and ends with David Baltimore's workhorse of immunity: NF-κB.[18,19] The major cells of innate immunity—monocytes, dendritic cells and neutrophils—are chock-full of TLRs. When TLRs recognize a discrete pattern on a microbe, virus or foreign debris, they signal where contact has been made. As the phagocyte writhes in response to invaders, the cell remains properly oriented by its two paired centrioles—their built-in gyroscope. The centrioles, which are oriented at right angles to each other, have a unique microtubular architecture that permits its distant radiations to sense signals from any direction. The gyroscope keeps the cell on track as the gymnastics of digestion or secretion changes its sense of front and back.[20–22] By this means, TLRs dictate what Wirschaus called the "spatial pattern" of the cell.

## Seeing Pictures, Not Pixels

It's hard to speak of spatial patterns without referring to Gestalt psychology. The Gestalt psychologists argued that our vision of the "whole" is innate, that we possess built-in neural mechanisms that reduce complex images, sound or living things to simpler, more concise forms.[23,24] This notion unites Gestalt psychology with developmental biology. It's an idea of the German Enlightenment, dating from Goethe and Kant, that's alive and well today. Nüsslein-Volhard quoted Goethe in her Nobel lecture:

*All creatures develop according to eternal laws*
*And the earliest image is hidden in the most complex form.*[14]

Gestalt psychology began in the summer of 1910 when Max Wertheimer got off the train in Frankfurt. The 30-year-old Czech psychologist was on his way to a Rhineland vacation, when, according to his son, he experienced a "Eureka!" moment. Wertheimer asked himself why anyone and everyone who sees a linear series of blinking lights at a railway crossing sees them as movement.[25] As a philosophy student, he'd poured over the ancient dialectics of atomism versus holism, of innate versus learned perception, etc. etc. But he got off the train convinced that the flashing lights had shown him a basic principle of cognition in general. He later recalled:

> *Regardless of whether or not one believes that [visual] relation-*
> *ships depend upon past experience, the question remains . . .*
> *Do such relationships exhibit the operations of intrinsic laws*
> *or not, and if so, which laws? Such a question requires experi-*
> *mental inquiry and cannot be answered by the mere expression*
> *"past experience."*[26]

In Frankfurt, he bought a zoetrope, a child's toy based on a moving picture machine invented by Eadweard Muybridge two decades earlier. A strip of pictures is placed inside the machine, it's rotated and when viewed though slits on the side, the stationary images come alive as a single, moving picture. Wertheimer took the device to his hotel room where he made his own picture strips, not of real objects but of simple abstract lines. By adjusting their geometry, he worked out the minimal variation required to produce the illusion of motion. The effect is now known as "apparent movement"[27] and is the reason that millions of people today can gawk at animated avatars. We see pictures, not pixels.

Wertheimer stayed on in Frankfurt and landed a position at the new Psychological Institute. There he was able to make use of a fast-shutter projector, the formidable tachistoscope. This machine permits the investigator to flash images on and off a screen at precise intervals. Using as experimental subjects two younger colleagues, Wolfgang Köhler and Kurt Koffka, Wertheimer concluded that the apparent movement flashed on the screen is dictated not by individual elements but by the overall kinetics of their appearance: the Gestalt. We see the whole picture, rather than the sum of its parts, but only when the timing is right. By 1912 he'd published "Experimental Studies of the Perception of Movement," the seminal work of Gestalt psychology.[28]

In 1914, Wertheimer, Koffka and Köhler were separated by the Guns of August. Köhler was stationed in the Canary Islands during World War I and did experimental work on the local apes. He confirmed the Gestalt notion that perception of the field is innate rather than acquired.[29] After the war, Gestalt entered its golden age. Wertheimer had accepted a faculty appointment at the University of Berlin, and Köhler was appointed director of its Psychological Institute. The graduate program they established, which favored experimental over theoretical work, drew wide attention and a flock of bright graduate students; their new journal, called *Psychologische Forschung* (*Psychological Research*), became the house organ. In the 15 years between the Armistice and Hitler, the institute and the Gestalt psychologists started a movement; some even called it a counter-rebellion against behaviorism. It crossed the Atlantic, and its apostles included Rudolf Arnheim, Kurt Lewin, Wolfgang Metzger, Hans Wallach and Kurt Gottschaldt.[30]

But to those of us who spend their time looking at patterns of genes or 'omes on microchips, perhaps the most enchanting and important of the Gestalt papers is Wertheimer's "dot" essay.[26] It is filled with abstract patterns of geometric dots, jiggles and waves that give the work the look of a child's book of puzzles. But the puzzles make a very adult point. In one of Wertheimer's illustrations we see a series of vertical rows of alternating dots and circles. To its right, the adjacent figure uses the same elements, but we read this pattern as a series of horizontal rows of circles and dots. The same subunits are employed in both as in the first linear pattern. But our brain counts patterns, not dots; indeed the vertical and horizontal axes look the same to viewers who read left to right or vice versa. Wertheimer explained that we have an innate tendency to "constellate" to see as belonging together those elements that look alike: dots with dots, circles with circles. It's how we distinguish the signatures of TLRs or NF-kB on chips these days: by pattern recognition.

As the nationalist (read Nazi) movement spread to pose an ever more serious threat to the Weimar Republic, the Gestalt psychologists paid ever more frequent visits to the United States. Coast to coast, they challenged the ruling behaviorist school and made the occasional convert. In 1924, Koffka left Europe for the United States, where he taught at Smith. And when the brownshirts came to power in 1933, Lewin was able to get to Iowa and Cornell. Wertheimer came to the New School in New York, where Alvin Johnson had established his graduate faculty as a "University in Exile." Köhler remained in Berlin until 1935, was harassed for associating with "the Jew Wertheimer" and arrived in the United States in 1935 to a scholar's welcome at Swarthmore.[30]

But the temporal and spiritual home of the Gestalt movement was the New School first and last. With frequent lectures by Lewin, Arnheim and the founding triumvirate, the graduate faculty attracted students and American faculty members of the first rank.[31] The tradition has continued to the present. Not too long ago, a distinguished professor came from Sarah Lawrence to teach experimental psychology at the New School: Gertrude Baltimore, David's mother. Wertheimer would have been delighted: pattern recognition and NF-kB in one family. It probably comes from connecting the dots.

# Not by the Sword, but Disease: Doctor Howe and General Shinseki

*Our soldiers in the Army of the Potomac are dying at the rate of three and a half in a hundred yearly; and in the Army of the West at the rate of five in a hundred . . . twenty-seven whole regiments laid low in a year, not by the sword, but by disease!*

—Dr. Samuel Gridley Howe, 1861[1]

*If you don't like change, you're going to like irrelevance even less.*

—Gen. Eric Shinseki, 2002[2]

*Even as we stand here today, there are veterans who have worried about keeping their health care. . . . they deserve a smooth, error-free, no-fail benefits-assured transition into our ranks as veterans, and that is our responsibility. Not theirs.*

—Gen. Eric Shinseki, 2008[3]

## "There's a Lot of Amazing Things. . . ."

THE OBAMA ADMINISTRATION brought with it a noticeable change to the entry lobbies of VA hospitals all over the country. Large photo portraits, frugally framed, of George W. Bush and his secretary of veterans affairs, Lt. Gen. James Peake, were replaced by portraits of their successors, Barack Obama and Gen. Eric Shinseki. The outgoing president had been on the job, so to speak, for eight years, but Gen. Peake had been coping with his for a mere thirteen months.[4] Dr. Peake, a former Surgeon General of the Army, replaced Jim Nicholson, who had served as head of the VA for the previous two and a half years.[4] Mr. Nicholson, a Colorado real estate developer, had sterling credentials for the job of heading the VA. Himself

147

a veteran, he had served as chairman of the Republican National Committee during Mr. Bush's presidential election campaign and was appropriately rewarded by appointment as U.S. Ambassador to the Vatican.[5] In July 2007, Mr. Nicholson was forced to resign from his own short-lived reign at the VA[6] largely because of pressure from Congress and veterans groups who rightly complained that "The VA under Secretary Nicholson has been woefully unprepared for the influx of Iraq and Afghanistan veterans, consistently underestimating the number of new veterans who would seek care and failing to spend the money Congress allotted to treat mental health issues."[7] Secretary Nicholson resigned shortly after President Bush had visited the Washington VA Medical Center to praise both the secretary and the VA: "There's a lot of amazing things taking place right here in this facility."[8]

Amazing, indeed. The U.S. Department of Veterans Affairs functions as a part of a welfare state within a state. It maintains a health care system that critics might call "socialized medicine," were it not in service of the ex-military. The secretary presides over a single-payer health care service that in 2009 attended to 5.3 million veterans at more than 1,400 sites including hospitals, clinics, nursing homes, rehabilitation treatment programs and veterans' readjustment counseling centers.[9] But the head of the VA is responsible not only for supervising medical services but also for the conduct of medical education and biomedical research. Today, most of the 170 VA hospitals are affiliated with the 126 American medical schools, and yearly more than 31,000 medical residents and 16,000 medical students receive some of their clinical training in VA facilities.[9] The VA spends over $40 billion a year, and supports biomedical research to the tune of over $400 million a year. The latter figure is about the yearly average of dollars allotted to the National Institute of Arthritis, Musculoskeletal and Skin Diseases from 2000 through 2009.[10] Based on familiarity with both sources of funding, I'd judge research carried out under VA auspices easily on a par with that sponsored by the NIAMDS. Happily, the quality of research has remained practically stable, despite perturbations in medical care caused by those turnovers at the top.

What has not remained stable is the rest of the VA system, which clearly failed to anticipate the facilities and manpower required to cope with the new wave of veterans from the wars in Iraq and Afghanistan. Swamped and confused, Peake and Nicholson remained buoyant and unapologetic. A civil suit filed by an Iraq war veterans' organization in 2007 prompted the AP to evoke the Katrina disaster: "SUPPORTING THE TROOPS? HECKUVA JOB, NICHOLSON!"[7]

## "Let's Assume the World Is Linear . . ."

The advent of Gen. Shinseki has rectified some of those failures of anticipation. The general is best known for his response to a question by Sen. Carl Levin as to the troop levels required for victory and peacekeeping in postwar Iraq. Speaking in his role as chief of staff of the U.S. Army before the Armed Services Committee on February 25, 2003, the general replied: "Something on the order of several hundred thousand soldiers are probably . . . required."[11] He gave this estimate based on the experience of a wounded combat veteran who had risen to four-star rank, had led NATO's Peace Stabilization Force in Bosnia-Herzegovina and commanded both NATO's land forces and the U.S. Army in Europe. He went on to say:

> Well, let's assume the world is linear. If we required a certain amount of troops per 25,000 population in the Balkans, if the world is not radically different, something of the same extent is going to be needed in Iraq.[12]

His estimate, of course, was quite different from those voiced at the time by Defense Secretary Donald H. Rumsfeld and Deputy Secretary of Defense Paul Wolfowitz. Soon after, Shinseki retired, under pressure, it was assumed. Three years later Gen. John Abizaid, commander of U.S. Central Command and the chief architect of U.S. military strategy in Iraq, told the same Senate committee, "Gen. Shinseki was right."[13]

Eric Ken Shinseki was born in Lihue, Kauai, Hawaii, on November 28, 1942—a time when Japanese-Hawaiins were not exactly the most favorite folk. Bright, articulate and promising, he was appointed to West Point and within a few months of his graduation in 1965 was sent to Vietnam. There he commanded a tank squadron, was thrice wounded and had his foot blown off by a mine. Returning to the United States, he earned a master of arts degree in English literature from Duke University and taught English and American literature at West Point.[14] As one learns in the military: "Dolts may go there, but dolts don't teach there."

His experience in the long and sloppy peacekeeping operations in the Balkans made him wary of easy solutions, quick fixes and inflexible plans. His command principle, "If you don't like change, you're going to like irrelevance even less," should stand him in good stead as he head's the VA welfare state within a state.

## A Doctor Without Borders

The line of descent of our VA welfare state is usually traced to the formal founding of the Veterans Administration in 1930. But it really started with the Sanitary Commission of Civil War times. The commission, formed by presidential edict in 1861, included doctors, divines and philanthropic lay folk, mainly women. These volunteers took on "to do for the troops what government could not do"—that is, to check into unvaried or ill-cooked food, clothing, cooks, camping grounds, malaria, in fact everything connected with "the prevention of disease among volunteer soldiers not accustomed to the rigid regulations of the regular troops."[15] It became the civilian guarantor of preventive medicine on the Union side, providing money, food, nursing care, dispensaries and hospital ships—tasks too complex for the small cadre of regular Medical Corps officers who had their cots full of the wounded.

The VA of today can trace its roots to a letter written by President Lincoln to Gen. Winfield Scott in 1865 suggesting that the postwar task of the Sanitary Commission was to be "at all times ready to recognize the paramount claims of the soldiers of the nation, in the disposition of public trusts"[16]—that is, jobs, pensions and well-being.

Among those who founded the Sanitary Commission was Dr. Samuel Gridley Howe (1802–1875), a warrior/scholar who, a century and a half before Gen. Shinseki, carried the American flag to the borders of Russia and the Ottoman Empire. His career before the founding of the Sanitary Commission is astounding.

Dr. Howe is remembered these days mainly as the 19th century's leading educator of the blind—and as husband of Julia Ward Howe, whose "Battle Hymn of the Republic" was composed on the stationery of the Sanitary Commission. But Dr. Howe, a Harvard Medical School product, was more than an effective advocate for the blind.[17] He was perhaps the first doctor without borders.

Howe was a social radical, an abolitionist who risked his life in the cause of social justice with what T. W. Higginson described as a "constitutional love for freedom and for daring enterprises, taking more interest in action than in mere agitation." The actions he took were on behalf of Greek *vs.* Turk, Pole *vs.* Russian, Irish immigrant *vs.* Boston nativist, fugitive slave *vs.* slave owner, Abolitionist *vs.* Unionist—and finally North against South.[17] Dr. Howe was persuaded that dirt and disease were social ills. His idea of a medical career was literally a life in the trenches.

Early in his career, he traveled to Greece to fight for that nation's freedom from the Ottoman Empire.[18] On arrival, he enlisted in the Greek Army and for the next six years served both as a soldier and as chief surgeon to the hospital at Nauplion. In 1829, he established a medical center in Aegina and a school for the blind in Corinth. He wrote home from Greece in 1826:

> *The affairs of the country do not go on so well as could be wished. We could beat off the Turks, let them come on as thick as they pleased, but this season they have sent an army of disciplined Arabians from Egypt, before whom the Greeks cannot stand. . . . They are now besieging Missolonghi, a very important town (fortified mostly by poor Lord Byron), and if they take it things will go badly.*[19]

They went from bad to worse, then improved, but remained unresolved until Greece became independent in 1832. Long after the Greek revolution, and even after the Civil War in the United States, Howe continued to be active in Greek affairs. In 1866, he returned to Greece with his wife to organize support for the new uprising of the Cretans against Ottoman tyranny. What was missing in those days was NATO!

### To the Credit of the American Name

After the Greek campaigns of the 1820s, Howe established himself in Boston and pursued new ways to improve the education of the blind. These brought him overseas again, to Paris this time, ostensibly to study the causes of optic degeneration. During the July revolution of 1830, he cast in his lot with the Marquis de Lafayette, a move that almost cost him his life a year later. At Lafayette's prompting, Howe became chairman of the American-Polish Committee at Paris. This crew included not only the Marquis and Dr. Howe but also James Fenimore Cooper (of those Mohican tales) and Samuel F. B. Morse (of the telegraph). Their aim was to provide relief supplies to the Polish political refugees who had escaped to camps in Prussia from the second Polish-Russian war. (Paging NATO again!)

Dr. Howe undertook to distribute the supplies and funds personally, but while in Berlin he had fallen afoul of the Prussians, who placed him in solitary confinement. Howe wrote to his Parisian colleagues from prison:

> *I'll be cool then and let you know where and how I am— snug enough, between four granite walls, in a wee bit cell, fast*

*barred and bolted, and writing by the light which comes in from a little grated window, or air-hole, eight feet from the floor. I am kept in perfect seclusion; not a newspaper is allowed . . . not a sound disturbs my meditations, save the sentinel's heels as he paces up and down the corridor.* [19]

He obtained some German works on the education of the blind—"I did not know of their existence in France," he wrote from prison. "I hope if pen and paper are granted me here to translate some good things. If by the next packet you hear not of my liberation, then do all that can be done for me." [19] Fortunately, his committee in Paris prevailed on the State Department to mobilize the Prussians into releasing Howe. The doctor was transported to the nearest border in a sealed post wagon, accompanied by two policemen. Jolting over rough byroads, Howe "suffered the torments of the damned . . . until a copious vomiting relieved my pains. I could not persuade them to get me a glass of water. My strength of constitution however enabled me to undergo the journey, which lasted six days and during which I was subjected to a thousand vexations." [19]

On his return to Paris, Lafayette wrote Howe: "I warmly feel it is the duty of the American Committee in Paris to offer you a vote of thanks for the manner in which our instructions have been understood and executed to the great comfort of the Polish soldiers, to the Credit of the American name and to the gratification of every good Heart and sound Mind." A grateful Howe recalled that he was even more pleased by the cries of *"Vive l'Amérique! and Vive la France!"* that he heard from the exiled Poles on his rattling journeys over Prussian roads. [19]

The Prussians had neither officially arrested Howe nor preferred charges; they offered no apologies. Many years later, however, when the King of Prussia gave him a gold medal "for philanthropic achievements in teaching the blind," Howe had the curiosity to weigh it and found that its value, in money, was equal to the sum which he had been forced to pay the Prussians for his prison board and lodging in 1832.

In Boston, Howe became perhaps the first American physician to interest himself in the study of what we would nowadays call "special education." The first school for the blind, the Perkins Institute, of Watertown, Massachusetts, was his creation, and Laura Bridgman (the Helen Keller of the abolitionists) was his student. But his work with the blind never preempted his more general social concerns. [17]

His commitment to the abolitionist movement culminated on September 5, 1861, when—in the words of the minutes—"A meeting called for this day was held at Dr. Howe's room, 20 Bromfield Street, to take into consideration measures tending to the Emancipation of the Slaves as a War Policy."[17] Emancipation went hand in hand with the sanitary movement. Later in 1861, he wrote from the field with the Sanitary Commission:

> *Our soldiers in the Army of the Potomac are dying at the rate of three and a half in a hundred yearly; and in the Army of the West at the rate of five in a hundred . . . pardon this outburst; but I lose patience at the delay to strike a righteous and killing blow into the very stomach of the rebellion by proclaiming emancipation under the war power, and enforcing it as fast and as far as we can; since every week's delay costs five hundred lives, and every month's two thousand.*[1]

The Sanitary Commission was the first of what are now called NGOs—nongovernmental agencies such as Doctors Without Borders—that render humanitarian services in times of war or tribal conflict. Indeed, the commission's work extended to healing the longer lasting wounds of battle—as in those 170 hospitals of the VA.

Doctor Howe would have been gratified by those new portraits of President Obama and Gen. Shinseki hanging on the lobby walls of our VA hospitals. For two of the doctor's lifelong hopes for America have been realized by a later generation. The election of Barack Obama shows us that we all gain when we all are equal, while General Shinseki has shown us that keeping the peace is the ultimate victory in war.

# Science as Oath and Testimony: Joshua Lederberg

*The demand for the meticulous prediction of which experiments will be done tomorrow, the detailed protocol of grant applications . . . flies in the face of scientific discovery, which is full of false starts, the serendipidy, the unpredictability of any great discovery or any real important consequence.*

—Joshua Lederberg, Stanford, 1978[1]

*Publication . . . converts private to public knowledge, in the service of registering a private claim of original authorship—in science, of discovery. Above all, the act of publication is an inscription under oath, a testimony.*

—Joshua Lederberg, Marine Biological Laboratory, 1991[2]

## Startled and Privileged

JOSH LEDERBERG WAS ALREADY WORRIED a generation ago: "One of the major trends of scientific writings for the past century is the systematic falsification of the actual techniques and method of discovery."[1] Lederberg wasn't concerned that published papers fail to mention the false starts, wrong turns or dead ends on the road to discovery. He accepted the conventional forms of scientific publication, which he called "recipes for replication of the results." He even praised the traditional format of the standard research report for its "pedagogical elegance" that prevented dragging in "all the dirty linen that led to the very fine fabrics that are eventually produced." What troubled Lederberg most, however, was that when we falsify the process of discovery we lead the gatekeepers of science astray. All of us are guilty, he argued, of persuading funding agencies to

154

expect grant proposals neatly packaged as "recipes for replication" with no wiggle room for risk and innovation. He warned that government or private entities that required tidy recipes for proposed experiments would be "selecting against creativity" in research.[1]

Joshua Lederberg used the evolutionary term "selection" accurately and with intent. The discussion took place at a 1978 Stanford symposium honoring the centenary of Claude Bernard's death; it was cosponsored by the French consul in San Francisco and took place at the Moët estate in the wine country. As expected, talk flowed as freely as vintage and soon the discussion turned to whether Bernard had learned anything from Darwin. A colleague suggested that a true scientific discovery is like a mutation, a quantum shift with selective advantage. To extend the analogy, Lederberg was asked whether system (pure theory) or experiment (the lab observation) was the better mutagen. His answer: both. He was convinced that the "tension between system and experiment" directs the sort of sea change in science that Thomas Kuhn was to call a paradigm shift.[3]

Joshua Lederberg's own career illustrates the point. Discussing his own paradigm shift in a joint interview with Thomas Kuhn, he attributed it to a mixture of system and experiment:

> *I was startled—and privileged—at age 21 to have made a surprising discovery that involved merging bacteriology and genetics. That was contrary to the wisdom of the time, which held that bacteria could not be crossed since they had no genetics. I've been puzzling about that ever since, because I felt the discovery should have been made 20 years before I was born.*[4]

Perhaps the discovery could have been made in 1905, but not in the United States, and not by someone from Joshua Lederberg's background.

## The Impediments of Personality and Race

In the half century between Victory in Europe and the Mess in Mesopotamia, American biomedical science underwent a radical social transformation. Before World War II, the United States had made a respectable showing in science on the world stage, but we did not attain a pre-eminent role until real and perceived national needs (like the Hitler War, the Cold War, Star Wars, Bio terror) made resources available to pursue basic science in pursuit of practical goals. Programs such as the navy's V-12 plan, the GI Bill, the establishment of the NIH, the "War on Cancer," the Apollo program and so on made jobs, money and facilities available countrywide for a generation of native and émigré scientists. Perhaps the best and the brightest of that generation was Joshua Lederberg; his threefold achievements as scientist, public citizen and communicator made him its spokesman.

Son of a rabbi, Joshua Lederberg attended Stuyvesant High School in Manhattan, one of the elite institutions established by Mayor Fiorello La Guardia to help sons and daughters of immigrants compete with graduates of posh private schools for admission to selective colleges. He entered Columbia College at the onset of the war in 1941 and enlisted in the naval V-12 program.[5] The navy plan condensed premedical and medical studies into five or six years with the aim of producing naval officers in quantity. In his sophomore year he began working part time in the laboratory of Francis J. Ryan on *Neurospora* while fulfilling his naval duties by working in a clinical lab at St. Alban's Naval Hospital. In June of 1946, Lederberg produced a respectable, if not astounding paper on back-crosses of *Neurospora* mutants with "Frannie" as first author.[6]

That same year, and before completing his medical education at Columbia, he moved to Yale to work with Ed Tatum, who had been Ryan's mentor. This callow Columbia predoc published an astonishing and prescient paper that attracted as little attention at the time as Mendel's work on *Pisum sativum*. His single-authored letter in *Science* carried the presumptuous title of "A Nutritional Concept of Cancer."[7] Based on the example of *Neurospora* adaptation to selective nutritional media, he made the daring suggestion that human cancer is due to somatic mutation:

> *The Neurospora experiments suggest a mutational origin [of nutritional adaptation] and that virus infection could play a corresponding role. A consequence of this simple concept is that cancer cells may be found to differ in their growth factor requirements from cells of normal origin when grown in vitro.*[7]

The passage anticipates not only the somatic mutation theory of oncogenesis but also the possibility of directed mutation (that is, by a virus). Lederberg did not go on to address the problem of human cancer at Yale. He was excited by the finding in 1944 by Oswald Avery, Maclyn McCarty and Colin MacLeod that DNA, and not protein, was the transforming principle in *Pneumococcus*[8] and moved from *Neurospora* to more tractable *E. coli*. He reasoned that if genes were made of pure DNA that could pass from microbe to microbe, then perhaps bacterial inheritance could follow Mendelian laws. Two centuries earlier, Linnaeus had shown sexual reproduction in plants—why couldn't this happen in bacteria? In Tatum's lab at Yale, Lederberg studied nutritionally adapted strains of K 12 *E. coli*—a lucky shot—and the results came fast. By June 1947, they reported the discovery that led them to a podium in Stockholm a decade later:

> *The conception that bacteria have no sexual mode of reproduction is widely entertained. This paper will be devoted to the presentation of evidence for the occurrence in a bacterium of a process of gene recombination, from which the existence of a sexual stage may be inferred.*[9]

Lederberg and Tatum had tested the common wisdom that bacterial reproduction was by clonal division and found it wanting.

On the basis of this and related work, Lederberg was awarded a Yale PhD in 1947 and was then faced with the problem of finding a job in academia. No luck at Yale or Columbia. Looking back, Lederberg recalled the obstacles of the day to an academic appointment of "a brash New Yorker, and a Hebraic one at that." Tatum wrote a number of letters of recommendation for young Josh to universities great and small: "Tatum took pains to argue that my research qualifications far outweighed the impediments of . . . personality and race."[10]

## Plasmids in Madison

Eventually Lederberg received a job offer from the University of Wisconsin, Madison, for an assistant professorship of genetics. Internal records of the time show a faculty divided over the offer: "Lederberg's background was metropolitan [and] weeks passed before a consensus was reached with reference to inviting Lederberg to Wisconsin."[11] One might note that in the immediate postwar period, most midwestern and a few southern universities were far more hospitable to people of a "metropolitan" background than schools such as Yale or Columbia: the names of Salvador Luria, Max Delbrück, Rita Levi-Montalcini, Sol Spiegelman and Henry Mahler come to mind. So, westward the course of Joshua.

For a dozen years, Lederberg flourished in the productive research environment of Madison: they were to be his golden years of discovery. Indeed, of his most cited papers, nine of ten date from the Madison years. With coworkers that included his first wife, Esther Zimmer, he described lysogeny and lambda phage. They also found that small circular runs of DNA, which he called plasmids, are distinct from chromosomal DNA and can undergo autonomous replication.[12] He devised novel techniques in bacterial genetics used the world over, made the beta-galactosidase of

*E. coli* a fit subject for analysis and introduced penicillin as a means of selecting protoplasts. In 1952, with his student and life-long friend Norton Zinder, he uncovered a third mode of gene transfer in bacteria.[13] The first, of course, was transformation (Avery, McCarty and MacLeod), the second was bacterial mating, which he and Tatum had discovered,[9] and the third was the viral transduction of Zinder and Lederberg, which they called "transduction." The discovery of transduction[13] fulfilled his prophecy in the 1946 *Science* paper: viral sequences can be inserted at will into foreign genomes to produce heritable change. Transduction and plasmid exchange have become the basis of modern biotechnology.

The Madison era was capped by the Nobel Prize in 1958, when at the age of 33 he shared the laurels with Edward Tatum and George Beadle. His banquet speech reflects his lifelong personal modesty: "Pride is humbled as humility is exalted in the dignity and splendor of this occasion."[14]

### Citizen of the Cosmos

At the height of this research flurry, Lederberg turned his attention to protecting the planets. He had already been active on a number of government advisory panels and commissions, but the intersection of space and microbiology presented him with a new career. The Russians had launched the satellite Sputnik in October of 1957, and Lederberg immediately understood that the microbes of earth and the heavenly bodies might cross-contaminate each other. Elected to the U.S. Academy of Sciences in the spring 1957, he wrote a letter to the academy in December warning them of this potential problem. Its solution required establishing a new field of study and novel research: the field acquired a new name, which Lederberg dubbed "exobiology" and its study has funded basic science for half a century. It was also due to his service on NASA and Academy committees on space biology that manned and unmanned missions were quarantined and decontaminated at each end of the flight: he became a founder member of NASA's space science board in 1958.[5]

In 1959, the young Nobel laureate moved to the sparkling new campus at Stanford's School of Medicine, where he became the first chair of the Department of Medical Genetics. The medical school soon learned that microbiology and human genetics were part and parcel of the same universe. Josh went J. B. S. Haldane one better in his approach to the host/parasite issue:

> *Haldane's most pungent remark was, "It is much easier for a*
> *mouse to get a set of genes which enable it to resist* Bacillus

typhimurium *than a set which enable it to resist cats." That may well be; he overlooks the unmatched evolutionary potential of the bacilli, which guarantees this will be an unending contest.*[15]

While his bench research moved on apace, he turned his attention to information technology with the establishment of the Instrumentation Research Laboratory. In 1964, together with computer scientist Ed Feigenbaum and polymorphic Carl Djerassi, Lederberg developed Dendritic Algorithm (DENDRAL), an artificial intelligence system that sought to introduce inductive reasoning into chemical analysis. He was also responsible for Stanford University Medical Experimental Computer for Artificial Intelligence in Medicine (SUMEX-AIM), a nationwide time-share computer network for collaborative genomic and other biochemical research projects. These efforts brought together experts in widely different fields: social scientists with physicians, biochemists with computer mavens, engineers with astronauts, soil scientists with rocketeers. And when in 1976 the first Viking spacecraft sampled the soil of Mars, the instruments used for chemical analyses were those of Lederberg's Instrumentation Lab.[5]

Public service brought Lederberg frequently to Washington: he served on President Kennedy's Panel on Mental Retardation and was chairman of President Carter's Cancer Panel. He consulted for the U.S. Arms Control and Disarmament Agency during negotiations for the Biological Weapons Convention in Geneva. And from 1966 to 1971, he published a weekly column, "Science and Man," in *The Washington Post,* which addressed problems that ranged from germ warfare to civil liberty, from the Middle East to the middle of the earth. They were "metropolitan" in outlook: liberal, secular and humane. He urged scientists to enter the public arena, avoiding the pressures of narrow professionalism:

> *From the perspective of my own participation in science, I certainly would not tax my colleagues with indifference to human problems. However, I believe that many of them are easily discouraged by larger problems and neglect to search for the ways in which their own expertise might be a unique key to solving a small problem, or perhaps more often to discovering an insidious new one.*[16]

Lederberg returned to New York in 1978 as president of Rockefeller University and set that eminent academy on an even steeper climb to excellence. While in New York, he became ever more interested in issues of national security: enrolled as a member of the U.S. Defense Science Board, he advised the government on defenses against bioterrorism, the control of biological weapons and—in keeping with his fear of cant—debunking the military equivalent of fibromylagia, the "Gulf War Syndrome."[5]

## The Imprimatur

The New York climate of intense intellectual exchange reinforced three of Joshua Lederberg's lifelong convictions: he believed in the importance of pure excellence in science, no matter what; he believed in the need for quantitative measures of excellence; and he recognized an obligation to make room for risk to achieve it. Perhaps such principles are not unexpected in a graduate of Stuyvesant High School—and a Nobel laureate, to boot.

Personally unassuming, warm and generous, he was tough as nails when it came to fashionable cant. He praised meritocracy and defended it against the political, financial and hierarchical enemies of promise. He asked a critical question in one of his early *Washington Post* essays:

> *Not long ago, I received an incredible demand, the more so as it was a formal requirement under United States law. It would compel me to look again at my colleagues and the staff of our university department with the eyes of a bigot to produce a racial census of employees belonging to certain minority groups. The purpose—to help enforce laws that forbid racial bias in employment on Government-aided projects—may be laudable. It is not that purpose but the means, namely calculated racial discrimination, that deserves critical discussion, not only because of its flimsy basis in scientific biology, but more importantly because it is setting the precedents for the kind of society we are building.*[17]

Joshua Lederberg founded two very successful, ongoing programs in the development of scientific careers, the Pew Scholars in Biomedical Sciences[18] and the Ellison Medical Foundation.[19] His one charge to the selection committees was that each applicant be judged not by geographic, gender or pigmentary criteria but by the answer to a simple question: "What's the discovery?" He also ruled that the proposals be short, innovative and free of the extensive data-dumps demanded by the NIH.

Lederberg was also a major proponent of quantifying any individual contribution to science, over and above its word-of-mouth reputation. He'd

had enough of the days of "personality and race." Prompted by Eugene Garfield's 1955 proposal for a "Scientific Citation Index,"[20] Lederberg soon
sent Garfield a supportive letter and promptly a Genetics Citation index
was underway. While one or another traditional critic carped at the value
of citation rankings, Lederberg and Garfield soon won out: it's now the
universally consulted Web of Science. Citation indexing has proved invalu-
able in sorting out the bloodlines of current research and its utility for the
republic of letters. Most recently, when the "h factor" was introduced as a
means of judging the effect of individual scientists,[21] Josh confided that he
was pleased to have come out so well on that score for work done so many
years ago.[22]

Lederberg was also convinced that the editing and review of scientific
publications was the best means of keeping excellence alive. In 1991 he
addressed an international conference of scientific editors at the Marine
Biological Laboratory in Woods Hole. His essay spelled out why, how and
what should appear in our literature:

> *Above all, the act of publication is an inscription under oath, a
> testimony. . . . I only need to remind you of the term "imprima-
> tur" (a wonderful metaphor): the imprinted witness that, an
> article having appeared in a refereed journal, it had survived
> a critical process, a conspiracy if you like, of the editors and the
> publishers and the referees. It is the essential ingredient to make
> scientific work responsible in the sense that one cannot readily
> retreat from assertions that have been signed, delivered to the
> printer and made available to thousands.*[2]

It's a fine standard, and we might say that it constitutes as much of Joshua Lederberg's legacy as his discoveries of viral transduction or bacterial sex.
*Resquiescat in Pace.*

# X-Ray Politics:
# The Nazi War on Röntgen and Einstein

*No light could come from the tube because the shield which covered it was impervious to any light known, even that of the electric arc. . . . I did not think. I investigated.*

—Wilhelm Konrad Röntgen, 1896[1]

*German physics? one asks. I might rather have said Aryan physics or the physics of the Nordic species of man . . . But, I shall be answered, "Science is and remains international." It is false. Science, like every other human product, is racial and conditioned by blood.*

—Philipp Lenard, "German Physics," 1936[2]

*People who could not spell the word "vote" or say it in English put a committed socialist ideologue in the White House. His name is Barack Hussein Obama. The revolution has come. It was led by the cult of multiculturalism aided by leftist liberals all over who don't have the same ideas about America as we do.*

—former Congressman Tom Tancredo, 2010[3]

*Nationalism is an infantile disease. It is the measles of mankind.*

—Albert Einstein, 1933[4]

## Judgment at Messelhausen

IT'S NOT OFTEN THAT AN ARTICLE in an old radiology journal evokes the stuff of movie drama. It's even rarer when it touches a raw nerve in the body politic of today. But an article in the August 1946 issue of *American Journal*

162

*of Roentgenology and Radium Therapy* reminds me of Burt Lancaster in the dock in *Judgment at Nuremberg,* a film based on the Nazi war criminal trials.[5] It also reminds me of the nasty, nativist sentiments bandied about by ultra-patriots in our own country today.

The radiology article casts a real-life American doctor, Lt. Col. Lewis E. Etter USAMC, in the role of an army prosecutor (Richard Widmark in the film) out to show what the Nazis were all about. In September 1945, a month after V-E day, Colonel Etter conducted two interviews with the wizened dean of "Aryan physics," Philipp Lenard.[6] Dodging the entry of American troops in March, the 85-year-old Lenard had fled from Heidelberg to Messelhausen, a tiny village in Bavaria. With Munich in ruins and Dachau exposed, Lenard's main worry was that the American occupation had slowed republication of his text. "But alien conquerors [*die Eroberer*] have now deprived Gutenberg's fine invention completely of all nobler uses."[7]

Philipp Lenard (1862–1947) was born in Hungary, where he was known as Lénárd Fülöp Eduárd Antal. Like other folks—then and now—who have been raised far from the center of national life, he became an ultranationalist. He had won the Nobel Prize in 1905 for extending Johann Wilhelm Hittorf's 1869 work on cathode rays and spent a lifetime knocking the work of Jewish, French and Polish Nobelists (Albert Einstein, Jean Baptiste Perrin and Marie Curie). He went so far as to write a proud history of German science, *Deutsche Physik,* that omitted Röntgen and Einstein entirely.[2] Not to include Einstein? No surprise. But Röntgen, the man who had won the very first Nobel Prize in Physics in 1901? There must have been another reason. "Was Röntgen Jewish?" Etter asked. Lenard replied "No, but he was a friend of Jews and acted like one."[6]

The Lenard-Röntgen story is the tale Etter was out to unravel. Had not Röntgen used a cathode ray tube invented by Lenard to get that image of his wife's hand: the first X-ray photograph? Colonel Etter, a future professor of radiology at Pittsburgh, prefaces his account with "There has been considerable controversy, with the help of the Nazi press and some party members over the position of Röntgen in science." The passage refers to Lenard's early Nazi Party membership and that Lenard, not Röntgen, was credited by the Third Reich with the discovery of X-rays. Etter elicited this astonishing claim from Lenard:

> *I am the mother of X-rays, Just as the mid-wife is not responsible for the mechanism of birth, so was Röntgen not responsible for the discovery of x-rays since all the groundwork had been*

*prepared by me. Without me, the name of Röntgen would be unknown today.*[6]

## The Smoking Gun

Pursuing this point, Etter confronted Lenard with several letters found in a safe at the Physics Institute in nearby Würzburg, an institute founded by Röntgen. The letters were hidden there under the terms of Röntgen's will. In the early 1920s, the National Socialists were already beating the drums of prejudice: it was clear to Röntgen that the nationalists were out to get him and to support one of their own, Philipp Lenard, in his claim to be the "mother of X-rays." In the summer of 1922 Lenard had written to the Nobel committee that Einstein was a "publicity-seeking Jew whose approach was alien to the true nature of German physics."[7]

Etter's article reprints photocopies of letters that show Lenard as a fawning opportunist flattering a recognized authority; he addresses Röntgen as a *Hochgehaltener Herr Professor* (Highly Esteemed Professor). The game is given away in a letter to Röntgen, dated "Heidelberg 21 May 97":

> *Because your great discovery caused such swift attention in the farthest circles, my modest work also came into the limelight, which was of particular luck for me, and I am doubly glad to have had your friendly participation.*[6]

In his Messelhausen interview, Lenard tells Colonel Etter that he had expected a reply from Röntgen "confessing" that X-rays were really their joint discovery: he couldn't have found them without Lenard's cathode ray tube. Etter discounts that explanation and dismisses Lenard's claims. Etter had been scouting around Würzburg and environs, aided in his quest by Dr. Harms, director of the institute, and by two non-Aryan GIs of the 115th Field Hospital, Bernard Berger and Joel Schwab, who served as interpreters during the interviews. With their help, Etter found physical evidence in Würzburg that Röntgen had not relied on Lenard's aluminum window tube (1893) in the critical experiments. Instead, he had used the earlier vacuum tubes devised by Hittorf and William Crookes (1869–1875).[6]

The Nobel committee agreed: on the evening of November 8, 1895, Röntgen took a Hittorf cathode ray tube covered with black cardboard to prevent light from entering. At some distance from his tube was a sheet of paper treated with the barium platinum-cyanide, ready for use as a screen in later experiments. As the current sparked in the vacuum tube, he noted that the paper was glowing in his darkened lab. He had moved it as far as six feet from the tube, much further than Lenard thought of placing a like target in his experiments. Röntgen realized that rays escaping from the Hittorf tube could not be the short-range rays that Lenard had explored but a new kind of emanation, "X-rays." These rays could penetrate human tissue and leave images on photographic films.[8] Etter explains that when Lenard missed emanations from his tube at distances greater than 8cm, "he missed the discovery of X-rays."

### They've Even Taken His Violin

Etter concluded that after Röntgen's death Lenard's animosity to Röntgen's reputation increased and "climaxed during the time of his lofty and commanding position in the Nazi hierarchy of scientists."[6] Lenard, an early party member, was the Nazi's point man in the purge of non-Aryan scientists from German universities. He had a trial run for his place in the hierarchy of Nazi science over a decade before Hitler came to power. After a 1920 physics meeting in Bad Nauheim, when he instigated a vigorous anti-Einstein campaign, first against Einstein's theories *(Einsteinismus),* then against his person. The campaign escalated to death threats, mobs of brownshirts around Einstein's home and violent disruptions of his lectures. The demagogue Rudolph Leibus offered a reward to anyone who would assassinate the hated "Jew scientist."[9]

By 1929, Lenard and his gang issued an antirelativity book, *One Hundred Authors Against Einstein,*[10] on which Einstein later commented, "If I were wrong, one would have been enough!"[11] But by then it was 1933, and Einstein had fled to England where he had been offered refuge in a vacation cottage on the Norfolk broads. The Conservative MP Godfrey Locker-Lamson, who owned the cottage, told Parliament: "Germany has turned out her most glorious citizen, Albert Einstein. The Huns have stolen his savings, plundered his place of residence, and even taken his violin. . . . How proud this country must be to have offered him shelter." But the Nazi threat unleashed by Lenard was felt all the way to East Anglia, and Einstein remained "guarded by two women secretaries, a farm laborer armed with rifles, and two detectives who questioned all visitors," until he was free to sail to America.[12]

In Germany, Joseph Göbbels, the Nazi propaganda minister, exulted as a crowd of Hitler youths burned Einstein's books on the square before Humboldt University: "The age of an overly refined Jewish intellectualism has come to an end, and the German Revolution has made the road clear again for the German character."[13] (Remind anyone of Tom Tancredo?) Be that as it may, the way was now clear for Lenard to replace Einstein as his country's most glorious scientist. And, sure enough, at a 1936 Nuremberg party rally attended by Hitler and Göbbels, Lenard received the National Science Prize from Hauptwissenschafftler (Chief Philosopher) Alfred Rosenberg:

> As a thinker Professor Lenard taught us that science is not equal to science, that racial souls alien to each other create quite different scientific worlds.[13]

### Roentgen Rays Without Borders

A decade later, Hauptwissenschafftler Rosenberg was hanged after judgment at Nuremberg. Philipp Lenard was excused by the Allies by virtue of his age and died in Messelhausen two years after war's end. Hitler and Göbbels met their end in a Berlin bunker.[14] And on a happier note, Einstein became an American sage, beloved by all at Princeton and beyond. Dr. Etter returned to Pittsburgh, where he became a respected professor of Roentgenology.[6]

And Röntgen? The affinity of "racial souls" had nothing to with the spread of X-ray science. The history of Röntgen rays provides firm evidence, instead, that science is indeed equal to science the world over. Within a year of its discovery, Röntgen rays went from lab bench to medical imaging; within a decade they were used to treat diseases from lupus to cancer.[15] In simple refutation of the Hauptwissenschafftler's racial notions, this rapid progress can be traced to scientists and clinicians in France and England; America and Japan; Jew and gentile; white and black; Slav, Pole and Magyar. Almost all of these breakthroughs were dependent on the general use of Hittorf/Crookes tubes—not those devised by the proud "mother of X-rays," Philipp Lenard.[16–24]

Röntgen's first record of his November X-ray discovery was published in the December 1895 *Proceedings of Physical-Medical Society of Würzburg*, a record of a lecture-demonstration to an appreciative hometown crowd.[17] A presentation in Berlin followed on January 4, 1896; two weeks later, Henri Poincaré who had received the news from Röntgen himself, immediately communicated its contents to the French Académie des Sciences. The Académie was only a small jump ahead of the developments overseas.

On January 31, the *New York Tribune* reported that in Berlin Röntgen's rays had shown a human hand and a compass in a metal case. The paper predicted that "In a similar manner a bullet in the human body light be located photographically; also the conditions of the internal organs could be photographed as an aid to medical diagnosis."[18] And by February 9 the paper reported that in New Jersey, "EDISON WILL TRY TO PHOTO-GRAPH A MAN'S BRAIN TOMORROW—THE APPLICATION OF THE ROENTGEN DISCOVERY TO MEDICINE REGARDED AS A MOST IMPORTANT FEATURE!"[19]

Two early applicants of Röntgen's discovery deserve particular mention. In June of 1896, six months after Röntgen's discovery, Leopold Freund (1868–1943) of the University of Vienna effected the first cure of skin cancer, in a five-year-old patient by means of X-rays generated by a Hittorf tube. He also wrote the first textbook in the field, but after the Nazis were welcomed to power in Austria, he was dismissed from his professorship and died miserably in exile. His country now recognizes that he "inaugurated radiotherapy as a new scientific specialty."[24] And then there's Georges Chicotot, a physician with credentials in both painting and Roentgenology. In 1910, he painted a self-portrait that documents a historic moment in radiotherapy: the first attempt to treat breast cancer with X-rays. In his left hand he holds a watch to time the exposure, in his right he holds a sort of extended Bunsen burner that spouts flames from its tip. He is heating the vessel that warms the vacuum tube. And that object glowing from the center of this painting? Well, it's a Crookes tube."[25]

Philipp Lenard was wrong on all counts. Science is and remains international, from Würzburg to Pittsburgh, Paris to Princeton, Vienna to Menlo Park. Let's try to keep it that way: a community of kindred spirits, rather than a pack of "racial souls."

*Acknowledgment:*

My attention was drawn to Dr. Etter's article by a good friend, Dr. Ervin G. Erdös (University of Illinois, Chicago).[26] He wrote that he heard about Etter's encounter with Lenard, "a flaming Nazi," from Etter himself when, as a colonel in the U.S. Army, "He opened a safe containing various documents. Among them were letters from Lenard to Professor Roentgen congratulating him for the discovery of X-ray imaging."[26]

# Wild Horses and
## *The Doctor's Dilemma*

Stillwater, OK: *Oklahoma State University administrators have quashed a project to test anthrax vaccines and treatment on baboons because the primates would have been euthanized. . . . Controversy about euthanizing animals after students performed surgeries on them arose after Madeleine Pickens, the wife of billionaire benefactor and OSU alumnus T. Boone Pickens, threatened to redirect a $5 million donation to the vet school because she did not agree with such practices.*

—Associated Press, November 30, 2009[1]

*You do not settle whether an experiment is justified or not by showing that it is of some use. The distinction is not between useful and useless experiments, but between barbarous and civilized behaviour. Vivisection is a social evil because if it advances human knowledge, it does so at the expense of human character.*

—George Bernard Shaw, 1909[2]

### Pros and Cons of Anthrax Research

ON NOVEMBER 30, 2009, *The Oklahoman* broke a news story that linked wild horses to yellow baboons and a halftime show at the Boone Pickens football stadium to anthrax research.[3] It seems that administrators at Oklahoma State University (OSU) had suddenly canceled their critical role in a $14.5 million NIH grant for studying the pathogenesis and treatment of anthrax in yellow baboons because the non-human primates involved in the research would be "euthanized."[3,4] OSU President Burns Hargis explained the official position:

*The OSU administration determined that this research was
not in the best interest of the university. The testing of lethal
pathogens on primates would be a new area for OSU that
is outside our current research programs. . . . Therefore, we
have decided not to pursue it . . . the administration has
simply decided that OSU will not have primates euthanized
on its campus. No animal rights advocates were consulted in
the process. The decision was reached after weighing all the
pros and cons.*[4]

Pros and cons? The "pros" of the issue are on record. The research proposed was a multiinstitutional study of anthrax pathophysiology and its prevention by vaccines in a primate model. The investigators are a team of well-regarded scientists, including Boston University's Shinichiro Kurosawa and K. Mark Coggeshall of the Oklahoma Medical Research Foundation. The baboon model they proposed had already resulted in a major new treatment option (protein C) for *Escherichia coli* infection.[5,6] Indeed, Kurosawa's earlier studies of anthrax in primates had documented changes in vascular permeability, disseminated intravascular coagulation and systemic inflammation, which closely mimicked those found in human patients. Studies in lower species were not as informative. The pilot study concluded that the baboon model "reveals that a fatal outcome is dominated by the host septic response, thereby providing important insights into approaches for treatment and prevention of anthrax in humans."[5] In the proposal squashed by OSU, Kurosawa et al. planned to test whether the current vaccine, the one now given to our troops, really worked in primates. They were first going to challenge baboons with spores of the attenuated Sterne strain of anthrax and then proceed to a study of the virulent Ames strain, which in 2001 infected at least 22 Americans and killed five. The vet school's Institutional Animal Care and Use Committee (IACUC) had unanimously approved the project, later noting that "euthanization" was routine in research protocols on experimental animals at OSU.[4] So much for the "pros."

## When the Donor's Name Is on the Stadium

The only "con" to the issue was identified immediately by local and national press reports and summed up in a blog on the website of *The Chronicle of Higher Education*: You don't cross the wife of a donor whose name is on the football stadium.[7] The president of OSU had it right; the research was certainly *not* in the best interest of the university. A few months before OSU canceled the anthrax project, Madeleine Pickens, a passionate animal rights activist and the fourth wife of investment mogul T. Boone Pickens, had diverted a promised $5 million gift to the veterinary school because the vet students performed "barbaric" surgery on dogs, which were then "euthanized."[8] These days, I've gathered, when people talk about "euthanizing" animals, what they really mean is "performing experiments on any animals at all." It's like using "intelligent design" when they mean "no Darwin at all."

More to this point, on November 19, 2009, while the OSU administration was still considering the pros and cons of primate anthrax research, Mrs. Pickens, founder of Saving America's Mustangs, brought her wild horse act into a facility named for her third husband, T. Boone Pickens. Mr. Pickens had given $165 million to the athletic department of his alma mater, and it surprised no one that the comely Mrs. Pickens, a former model, flight attendant and horse breeder, would host the pregame activities and halftime show at a game between the OSU Cowboys and the Buffaloes of Colorado. Featuring three wild mustangs—Freedom, El Compadre and Felio—the show attracted wide attention: "You can go to a town hall meeting and be lucky if 1,000 people show up," crowed Mrs. Pickens, "but if you go to a football game in a place like Stillwater, you'll have 50,000."[9] The Cowboys won 31–28.[10]

To be fair, the wild-horse campaign is by no means a pure exercise in animal sentimentalism. The U.S. Bureau of Land Management removes 6,000 to 10,000 wild burros and horses from the open range in ten Western states each year to maintain a stable population of about 27,000.[11] Hearing that some of the removed animals might be euthanized, Mrs. Pickens founded Saving America's Mustangs, whose members included former NFL stars Roger Staubach and Troy Aikman, as well as Jenny Craig (of the slimming company) and T. Boone himself, listed as "legendary business man." The group, featured on the madeleinepickens.com website, aims to establish a public-private sanctuary for the wild horses of the West by means of private donations and government action.[12] A skeptic might wonder whether it was simply by accident that a website of a large benefactor of OSU, dedicated to the welfare of wild horses, would salute the OSU decision on anthrax research in these terms: "Kudos for a Great Decision!—OSU President Cancels Anthrax Study Proposal Requiring Primate Euthanasia."[13] President Hargis could not have been displeased by praise from such quarters. It must have weighed heavily against protest from the ranks of experimental biology.[14] Sadly, as reported in *The Scientist,* the anthrax proposal was exactly the kind of study for which a new OSU Biosafety Level (BSL) 3 facility had been built. The new lab was intended not just for OSU researchers. In fact, the University of Oklahoma Health Sciences Center opted not to build a BSL 3 primate facility, as this one would be available at OSU.[4]

After weighing in all those pros and cons, the OSU administration had handed one more victory to animal sentimentalists over experimental biologists.

### Vivisection and Its Discontents

Opposition to animal experimentation has by no means been limited to the ranks of wealthy socialites or aging Brigitte Bardots. What is now a campaign against "animal euthanasia" began as a cry against "vivisection." The conflict has been ongoing since scientists in the 1830s first started poking into the live bodies of four-legged creatures. When François Magendie (1783–1855) opened the first lab of experimental physiology in Paris, his work immediately aroused both scientific and lay opposition. A preeminent natural philosopher, Baron Georges Cuvier (1769–1832), whose forte was comparative anatomy, distrusted work on living creatures, claiming that data thus derived lacked any theoretical basis.[15] Moreover, as pure chemistry was triumphant after M. E. Chevreul (1786–1889) and E. I. DuPont (1771–1834) established its use for textiles and munitions, chemists got the good labs, and the early physiologists—many of them only clinically trained—were relegated to dark basements. Vivisection also offended the fastidious laity. In the days before antisepsis and adequate anesthesia, the work was grubby, bloody and loud.

Claude Bernard (1813–1878), who learned the craft from Magendie in the 1840s, recalled that animal experiments were new and unpopular: "As soon as an experimental physiologist was discovered, he was denounced, became the abomination of his neighbors, and was handed over to the police for prosecution."[16] He knew whereof he spoke. Bernard's wife, Fanny, had become a devout, active member of the new antivivisectionist movement (Society for the Protection of Animals), founded by the influential Comte de Gramont. Fanny not only denounced her husband to the police for the "crime" of animal experimentation but also forever turned their two daughters against him. After 24 years of marriage, the mismatched couple separated. The break came shortly after Bernard published *Introduction à l'étude de la médecine expérimentale* in 1865, his plea for a new medicine based on new physiology: the dog lab.

Bernard turned his affections elsewhere. He befriended the scientifically curious, intellectually progressive Mme. Marie Raffalovich. She was the attractive wife of a well-off, complaisant banker; she was also a patron of the arts who ran an active salon. Bernard and Mme. Raffalovich exchanged over 500 letters over the next decade; his survived, and hers were destroyed.[17] The correspondence is a one-way record of their joint excursions into English, German and Russian scientific literature, of their broader interest in the visual arts, and the drafts of a new concept that Bernard had drawn from vivisection: the constancy of the internal environment.[16,17]

### From Claude Bernard to George Bernard Shaw

Here, the story of Claude Bernard, the vivisector who founded experimental medicine, runs into the work of George Bernard Shaw (1856–1950), the major pamphleteer of antivivisection in the 20th century. Mme. Raffalovich's son, Marc-André (1864–1934), was sent to Oxford for his education, became a poet and fell in with a fin-de-siècle, round-robin crowd of London dandies who included Oscar Wilde, Aubrey Beardsley, John Addington Symonds and poet John Gray.[18] Marc-André and John Gray became lovers and enjoyed a happy, lifelong civil union. The experience prompted Marc-André to write pioneering essays demanding equal treatment for what was then called a "Uranian" choice of partner.[19]

Two people from that group of Uranian dandies have been immortalized in English literature. Gray served as the model for Oscar Wilde's *The Picture of Dorian Gray*,[18] and George Bernard Shaw transformed Beardsley into the louche artist Louis Dubedat in *The Doctor's Dilemma* (1909).[20]

For reasons that might appall or delight Wilde and Shaw, they have been placed in the pantheon (and on the website) of the most radical of antivivisectionist groups of our day.[21] The Abolitionists—an animal rights group—would have us turn the clock back to the days before Magendie and Bernard. A "declaration of war" on those who perform experiments on animals is tempered with an offer of armistice on behalf of Wilde and Shaw:

> *Lay down your weapons. Lay down your scalpels and prods.*
> *Lay down your Pavlovian slings and restraint chairs. Lay*
> *down your stereotaxic devices and your rodent guillotines. Lay*
> *down your wires that shock and plates that burn. We welcome*

*you into our ranks—ranks that have known the likes of . . .*
*Oscar Wilde and George Bernard Shaw.*[21]

## The Doctor's Dilemma and Oklahoma State

The Abolitionist declaration is an echo of an argument advanced in Shaw's scathing preface to *The Doctor's Dilemma*:

> *Making all possible allowance for the effect of the brazen lying of the few men who bring a rush of despairing patients to their doors by professing in letters to the newspapers to have learnt from vivisection how to cure certain diseases, and the assurances of the sayers of smooth things that the practice is quite painless under the law, it is still difficult to find any civilized motive for an attitude by which the medical profession has everything to lose and nothing to gain.*[2]

Shaw played with these issues in *The Doctor's Dilemma*. In the play, a capable clinical investigator, Sir Colenso Ridgeon, has discovered a kind of BCG vaccine that will work, even on active tuberculosis. However, his supply is limited, and the doctor's dilemma is whether to give the last remaining doses to an aging, dull-but-honest fellow practitioner, Dr. Blenkinsop, or to a thoroughly unscrupulous, if talented, young artist, Louis Dubedat. The one denied has the misfortune to get the cheaper treatment, one that will "stimulate the phagocytes."

Shaw anticipated the modern dilemma of health care—public or private—in which the cost-benefit ratios of various treatments are tabulated, and care is denied those—the aged, for example—whose costs might sink the proverbial raft:

> RIDGEON. . . . . *I have at the hospital ten tuberculous patients whose lives I believe I can save.*
>
> MRS. DUBEDAT *(wife of Louis). Thank God!*
>
> RIDGEON. *Wait a moment. Try to think of those ten patients as ten shipwrecked men on a raft—a raft that is barely large enough to save them—that will not support one more. Another head bobs up through the waves at the side. Another man begs to be taken aboard. He implores the captain of the raft to save him. But the captain can only do that by pushing*

*one of his ten off the raft and drowning him to make room for the new comer. That is what you are asking me to do.*

MRS. DUBEDAT. *But how can that be? I don't understand. Surely—*

RIDGEON. . . . *The treatment is a new one. It takes time, means and skill; and there is not enough for another case. Our ten cases are already chosen cases. Do you understand what I mean by chosen?*

MRS. DUBEDAT. *Chosen. No: I can't understand.*[22]

Mrs. Dubedat's lack of understanding was Shaw's as well. By definition, antivivisectionists, or those who oppose euthanization, are *choosing* to keep future generations off of a raft on which they are already safe. The "raft" is that safety zone of public health, of vaccines, of antisepsis, antibiotics, insulin, cortisone and all the rest—a zone in which life expectancy doubled from age 40 in 1840 to approximately 80 by 1980. The raft has been constructed by generations of biomedical researchers. Many of us owe our place on that raft to the humane use of animals, including primates, in biological research. We need more people on the raft, not fewer, and the way to achieve this is spelled out in the closing paragraph of the FASEB Statement on the Cancellation of OSU Primate Anthrax Project:

> *While fiduciary responsibility to the institution is an essential part of a university president's function, equally imperative is the appreciation of the research mission of the university and protection of the intellectual pursuit and scientific principles. Animals play a critical role in research that helps save both human and animal lives, and their use is strictly regulated under a system of scientific and ethical oversight. Research institutions should be prepared and willing to defend both the principle and the system that contribute to the health of the nation.*[14]

# Glass Ceilings at the Nobel Prizes

*All my colleagues have told me that it is preferable you not come here on December 10. I therefore beg you to remain in France [and] hope that you will telegraph the Secretary of the Academy or even me that it is impossible to come.*

—Svante Arrhenius to Marie Curie, 1911[1]

*[Harvard President Lawrence] Summers offered three possible explanations, in declining order of importance, for the small number of women in high-level positions in science and engineering. The first was the reluctance or inability of women who have children to work 80-hour weeks. Summers [also] said that women do not have the same "innate ability" or "natural ability" as men in some fields.*

—*The Boston Globe*, January 17, 2005[2]

*That was the case two weeks ago, when Greider was up, as usual, before 5. With time to spare before going to spin class, she was folding laundry [when] the call came. Several days later, when she heard that President Obama had won the Nobel Peace Prize, she thought to herself: "I bet he wasn't folding laundry."*

—*The Washington Post*, October 20, 2009[3]

INTERVIEWER: *I just wanted to ask you [why] telomerase and telomere research is a field which has, happily, a large number of women working in it?*

ELIZABETH BLACKBURN: *I'll turn your comment around and say it's fairly close to the biological ratio of men and women. It's all the other fields that are aberrant.*

—Nobel Interview, October 5, 2009[4]

## The Stimulus of Fame

IN DECEMBER OF 2009, aberrancy ran into equality on a podium in Stockholm. Three of the six new Nobel laureates in fields related to experimental biology were women: Elizabeth Blackburn of the University of California San Francisco and Carol Greider of Johns Hopkins University in physiology or medicine and Ada E. Yonath of the Weizmann Institute of Science in chemistry. Adding the female laureates in literature and the Nobel Memorial Prize in Economics, the Nobel committee was proud to note that five of the 13 new laureates were women, the largest number ever to mount the podium in Stockholm. Blackburn and Greider had the distinction of breaking another glass ceiling: it was the first time that any Nobel Prize in the sciences had been awarded to more than one woman.[5]

The awards were no surprise: Blackburn, Greider and Jack Szostak, their fellow laureate, had received honors galore. In 1982, Blackburn and Szostak had shown that telomere DNA from Tetrahymena could protect chromosome shortening in yeast,[6] and on Christmas Day of 1984, Greider got the first inkling of telemorase, a reverse transcriptase that adds DNA to telomeres by means of an RNA template.[7] Telemorase solved the "end replication problem" by explaining how DNA could be added to one of the two strands of DNA as the polymerase comes to the ends of chromosomes during cell division.[8]

Ada Yonath became only the fourth woman to have won the Nobel Prize in chemistry—for her X-ray diffraction solution of ribosomal structure. She describes the results succinctly:

> *The ribosome is a machine that gets instructions from the genetic code and operates chemically in order to produce the product. . . . The product is a protein and if you think about the kangaroo in a pocket, the product goes first into a pocket which is actually in the ribosomal tunnel.*[9]

The marsupial analogy is correct and resulted from Yonath's application to crystals of bacterial ribosomes of the Bragg equation, $n\lambda = d\sin\theta$ (William and Lawrence Bragg, father and son, Prix Nobel, 1915). Protein crystallization is as much art as science, and crystallizing large molecules is the greatest art of all. It took 20 years, and once the art was accomplished, new science could begin. Sure enough, it turns out that the 3-D structure permitted Yonath and her colleagues to determine that many potent antibiotics work by preventing newborn proteins from leaving the kangaroo's pouch, the ribosomal tunnel.[10]

Yonath describes her admiration for Dorothy Crowfoot Hodgkin, also a crystallographer. Hodgkin, a mentor for other female crystallographers, such as Rosalind Franklin, was the third woman to win a Nobel Prize in chemistry (1964). She had worked out the 3-D structure of penicillin and vitamin $B_{12}$. "There was Marie Curie (1911) and her daughter (Irene Joliot-Curie, 1935) and . . . Dorothy Hodgkin and now it's me," said Yonath.[9] It's been a long time coming.

For centuries before Lawrence Summers's gaffe, women were considered intellectually inferior to men, especially in the "hard" sciences. Summers and his ilk had failed to ask a question posed by Wendell Philips in 1851 at the Second National Woman's Rights Convention in Worcester:

> When woman has enjoyed for as many centuries as we have the aid of books, the discipline of life, and the stimulus of fame, it will be time to begin the discussion of these questions: "What is the intellect of woman?" "Is it equal to that of man?"[11]

In 2010, the answer is to be found in every lab, on every lecture podium, and on every editorial masthead; equality is beating out aberrancy (Elizabeth Blackburn's term). The "stimulus of fame" has played no small part; Irène Joliot-Curie might testify to that. Indeed, the worldwide enthusiasm with which the 2009 awards have been greeted should assure us that many more Elizabeths, Carols, Adas and Irènes will mount the podium in Stockholm.

## It Is Not Good That Man Should Be Alone

It wasn't all open arms and worldwide celebration each time Marie Curie learned that she had been awarded a Nobel Prize. Her first (physics, 1903) had been with husband Pierre and with Henri Becquerel. In fact, her name had been added almost as an afterthought only after Pierre had intervened with the committee, writing that her inclusion would be *plus joli d'un point de vue artistique*" (more attractive from an esthetic point of view). The condescending note continued to the prize ceremony itself, when on December 10, 1903, Dr. H. R. Törnebladh, president of the Royal Swedish Academy of Sciences, quoted Genesis:

> The great success of Professor and Madame Curie is the best illustration of the old proverb, coninucta valent, in union here is strength. This makes us look at God's word in an entirely new light: "It is not good that the man should be alone; I will make him an helpmeet for him."

The helpmeet motif was the tune of the day, as newspaper and magazine images of the two winners always depicted the figure of a passive, seated lady receiving a dangerous object from an erect French male.

By 1911, when Marie Curie was nominated for her second Nobel Prize, fortune had turned against her. Pierre had died suddenly and tragically, she had lost a bitter election fight to the Académie des Sciences, and worse yet, she was involved in a duel-enlivened love triangle with Paul Langevin, a physicist colleague of Pierre's. Indeed, Marie Curie was in imminent danger of being involved in a trial for alienation of affection, scheduled for—of all days—the day of the Nobel Prize ceremony December 10, 1911. Her detractors had argued in the press that she was "an alien, a Polish woman, a researcher supported by our French scientists," who had "come and stolen an honest French woman's husband."[12]

The trial and attendant publicity prompted that letter from Arrhenius asking her not to attend the ceremonies.[1] Maria Salomea Sklodowska Curie replied to the committee that she had been given the prize for her discovery of radium and polonium and that she "could not accept the principle that appreciation of the value of scientific work should be influenced by slander concerning a researcher's private life."[12] In the event, the litigation was settled in the first days of December, and she traveled to Stockholm. At the ceremony, Dr. E. W. Dahlgren, president of the Royal Swedish Academy of Sciences, again stressed the male contribution to the discovery of radium and polonium:

> *We know that radium claims its most promising results especially in the treatment of cancerous growths and of lupus. . . . the Royal Academy of Sciences considers itself well justified in awarding the Nobel Prize for Chemistry to the sole survivor of the two scientists to whom we owe this discovery, to Mme. Marie Sklodowska Curie.*[13]

Marie Curie's reply was implicit in her lucid, self-confident acceptance speech. She spoke of "All the elements emitting such radiation *I have termed radioactive* [editor's italics], and the new property of matter revealed in this emission has thus received the name radioactivity" and "*I was struck by the fact* [editor's italics] that the activity of uranium and thorium compounds appears to be an atomic property of the elements." She clearly distinguished her own contribution from those made by both Curies working together.[14] It was a stimulus to fame that has outlived her detractors.

## The Sole Survivor

On April 19, 1906, the 47-year-old Pierre Curie was run over by an oversized, horse-drawn wagon filled with bales of army uniforms. He was negotiating that tricky Parisian intersection where traffic from the Rue Dauphine, the Quai Conti, the Quai des Grands Augustins and the Pont Neuf has created Gallic havoc for over a century. Curie had just left a meeting of reform-minded university professors, where he argued for legislation to improve the lot of junior faculty and to prevent laboratory accidents. He had planned to stop at his publisher's office on the Quai, but the office was shut because of a strike by equally reform-minded trade unionists. Absent-minded and somewhat radium-sick, he turned away in the spring rain and was on his way to the library of the *Institut* when that six-ton wagon rumbled down the bridge from the Île de la Cité to crush his skull.[15]

Marie Curie later recollected that on the Rue Dauphine, "I lost my beloved Pierre, and with him all hope and all support for the rest of my life." She was right; although Curie was to survive her husband until 1934, her contributions to science after 1911 were less focused on day-to-day laboratory work. She turned her tough mind to the application of her discoveries, to teaching young scientists and to construction of the Radium Institute, which she turned into a world center of physical science.

Curie's generous 1911 Nobel lecture spells out the details of an incredible run the two Curies had together. In the course of six short years, they had laid the foundations for the next century of physics and set the clock of our atomic age. Pierre had already become famous for his work with his brother, Jacques, on piezoelectricity (some crystals, such as ceramic or bone, generate an electric current when compressed). He had earned his doctorate for studies with Paul Langevin on paramagnetic resonance (the moment of an atom or electron varies inversely with the temperature). It was the same year (1895) that Wilhelm Röntgen took the first picture of the bones of his wife's hand by means of his novel rays.

By 1897, Henri Becquerel had found that uranium also produced rays—emanations—that left Röntgen-like shadows on photographic plates kept in the dark. Almost simultaneously, William Thomson, Lord Kelvin, discovered that the "ionizing" emanations from uranium imparted an electric charge to the air. In December of that year, Pierre and Marie set out to quantify the Becquerel emanations—ionizing radiation—of a great variety of natural substances. For this purpose, they used the piezoelectric quartz balance, an instrument that Pierre had designed, and by February had found that the residue of pitchblende, from which uranium had been extracted, gave far greater signals than uranium itself. They deduced correctly that there was an ionizing substance far more active than uranium lurking in the sticky brew. It was the same year that Émile Zola wrote *J'accuse*, and France split forever into the supporters of the falsely accused Alfred Dreyfus, the Dreyfusards, and their right-wing opponents.

By the end of 1898, the Curies had postulated that the new element, which she had dubbed "radium," decayed into another, which she called "polonium," after the country of her birth. "Radioactivity" was the new name for emanations from these elements.[14,16] In 1902, by means of heroic preparative procedures, Marie Curie, at last, isolated radium in pure form. Later that year, Pierre calculated that 1 g radium emitted $3.7 \times 10^{10}$ disintegrations per second; we call this amount of radioactivity one Curie. Shortly thereafter, Pierre made the heuristic discovery that 1 g radium could heat 1 g water from 0° to 100°; we call this sort of transformation "atomic energy," and nowadays it powers more than half of France. By 1903, the year that Pierre and Marie won the Nobel Prize, they had also come down with the first signs of radium sickness.

For six unmatched years of discovery in the setting of the Third Republic, axes were drawn between right and left, church and state, theory and application, and risk and benefit of a new science in a new century. It's a grand story, and although the Curies are on the spoor of the new, with the Dreyfus case breaking about them, it's an exemplary tale of science in service to reason. However, after Pierre's death on the Rue Dauphine, the story of Marie Curie becomes less of a life in science, as the outrageous attacks on her by the anti-Dreyfusard press turned her attention from science to the broader social scene.[17] Her public efforts proved to be as successful as her work in the lab. It was in recognition of the many mobile X-ray units she organized during the First World War that a grateful France forgave her for the Langevin affair by permitting her to establish the Radium Institute.

### The Republic Does Not Need Scientists

No published material explores what must have been the remarkable relationship between Madame Curie, a pale, intense widow in a plain black dress who lived on the fashionable Quai de Béthune, and her daughter, a physicist at her mother's institute, who married a brilliant young coworker. Playing out the story of *Marie et Pierre Redux*, Irène and Frédéric Joliot-Curie shared not only the Nobel Prize in chemistry for induced radioactivity in 1935—five prizes in one family—but also an abiding attachment to the Communist Party. As the Dreyfus case had been the cause that engaged Pierre and Marie, the Popular Front of the 1930s enlisted Irène and Frédéric. The story of the Curies reached from the Quai de Béthune to the ranks of the Comintern. Both generations encountered enmity of the most virulent sort from nativists and anti-intellectuals.[18]

In her losing battle for election to the Académie des Sciences in 1911, Marie had found staunch allies in her fellow Dreyfusards of the Sorbonne. Mathematicians, physicists and chemists, such as Paul Appel, Gabriel Lippmann and Henri Poincaré became the targets of the proto-fascist ranks of La France profonde. Léon Daudet also led the right-wing attack on Marie Curie's nomination for the 1911 Nobel Prize. Attacking the Godless Sorbonne professors in L'Action française, he accused Curie's champions of no longer "hiding behind the Lives of the Saints but behind algebra, physics and chemistry treatises."[19] He reverted to Antoine Quentin Fouquier-Tinville's notorious cry that had sent Lavoisier to the guillotine: "The Republic does not need any scientists."[12] Certainly, the Republic of Daudet did not need any women scientists. Daudet's mother, Julia, had cast a traditional curse at the likes of Marie and Irène Curie:

> Science is useless to women, unless they are the exceptions who are inclined to a masculine career, and that is always too bad . . . this excessive independence of ideas, quest for liberal ideas, usurpation and intrusion in the role of lawyer or of intern in the hospitals. . . . all that seems to me the fantasies and ambitions of those with dormant hearts, women without children or households.[20]

On December 10, 2009, five of 13 women Nobel laureates, distinguished for their independence of ideas and their quest for liberal rather than nativist values, received the plaudits of that distinguished assembly in Stockholm. Elizabeth Blackburn, representing a generation of scientists, who are also wives and mothers, had an answer for Mme. Daudet and her ilk: "Our lives were work and family and that was it. . . . that wasn't a sacrifice; we love both our family and our work."[21] One could hear a glass ceiling crack.

# Medea and the Microtubule

*To improve human health, scientific discoveries must be translated into practical applications. Such discoveries typically begin at "the bench" with basic research—in which scientists study disease at a molecular or cellular level—then progress to the clinical level, or the patient's "bedside." [But] translational research is really a two-way street in which clinical researchers make novel observations about the nature and progression of disease that often stimulate basic investigations.*

—NIH Roadmap for Medical Research, 2009[1]

*[Medicine]: from the Indo-European root MAD or MED to reflect, to think, to meditate on ideas; In Greek μηδεα or medéa meaning plans or counsels and subsequently schemes. This last meaning is seen in the Medea of Greek tragedy, Medea the Sorceress, the Schemer, the cunning.*

—Thelma Charen, "The Etymology of Medicine," 1951[2]

MEDEA: *The best way is the most direct, to use the skills I have by nature and poison them, destroy them with my drugs. I know the drugs required for such things.*

CHORUS: *Poor thing, the woman from Colchis, so unhappy . . .*

—Euripides, *Medea*, 431 BCE[3]

*Lord, do I have to listen to all this melodrama?*

—Tyler Perry's *Madea Goes to Jail* (film), 2009[4]

## Translating Little Science into Big Science

THESE DAYS, SUPPORT FOR "TRANSLATIONAL RESEARCH" is at its high-water mark. Three years ago, the NIH Roadmap for Medical Research modestly laid out a "bolder transforming vision for the 21st Century" as it launched its Clinical and Translational Science Awards (CTSA) Consortium.[1] By the end of 2009, the national consortium funded 39 centers in 23 states, and by 2012, the NIH expects to establish 60 centers nationwide, with an annual funding commitment of $500 million. On the crest of this wave, *Science* launched a new magazine in October 2009, *Science Translational Medicine*,[5] which competes with such stalwarts as the *American Journal of Translational Research* (established January 2009)[6] and the venerable *Journal of Translational Medicine* (established 2003).[7] A high tide lifts all boats: the long-running *Journal of Clinical and Laboratory Medicine* changed its moniker to *Translational Medicine* in 2006, the very year the NIH announced the CTSA.[8] Sure enough, the flag officers are coming on board. The American Association for the Advancement of Science (AAAS) and *Science*, in conjunction with leaders of a score of eminent societies and foundations, are launching the Clinical and Translational Science Network (CTSciNet), an online community that will combine a career-development web portal for clinical and translational investigators with an "experimental evolving communications infrastructure."[9]

What might account for this sudden burst of interest in translational research, a term traditionally applied to bringing science from the bench to the bedside? Many would argue that the last half century of individual discoveries in molecular and cell biology laid the groundwork: "little science." Just a short decade ago, two legions of researchers, public and private, solved the double-crostic of the human genome: "big science" at its biggest. That achievement, and promises made at the time (e.g., "The Language in Which God Created Life"), aroused high hopes among scientists and the public alike for a quick translation from divination into medical practice.[10] As expected, our leaders have responded by forming the most worthy Clinical and Translational Science Awards Consortium: "big bucks."

However, we'd do well to wait a bit before expecting big news from the bedside. What the group assault on the human genome has already accomplished, and in record time, I'd argue, is to prompt the new "'omic" revolution. "'Omics," FISH (fluorescent in situ hybridization) and chips, and systems biology have produced armories of new tools that now include proteomics, lipidomics, metabonomics, nutrigenomics and transcriptomics, *und so weiter* (big data).[11] With mastery acquired over these mountains of data by the revolution in information technology, the stage is set to plot abscissas of the bench against the ordinates of bedside with astounding results for both basic and clinical science.[12]

What about the other direction: bedside to bench? The flag officers who founded the Clinical and Translational Science Network correctly point out that

> *[Translational research] is not one-way; the insights gained at the bedside, and from clinical and population-based studies, will spawn hypotheses, enabling scientists to probe the mechanisms of disease in new ways and ultimately enriching basic biology.*[9]

Indeed, translations from the bedside to bench move more quickly than in the other direction. Many of us will remember that the first postmodern fashion in biomedical research was to label every possible discipline and every possible disease as "molecular." Dozens of journals sprouted in the last half of the last century with titles ranging from *Molecular Ecology* to *Clinical and Molecular Allergy*. The molecular revolution in biomedicine traces directly to Linus Pauling's analysis of sickle cell anemia as a "molecular disease." James B. Herrick first described sickle-cell anemia, with its characteristic deformity of red cells, in 1910; 39 years later Pauling looked at diffractions from those hemoglobin crystals in a Debye camera, and molecular biology was on its way.[13] A century after Herrick, Pauling's structural biology has moved from one laboratory feat to another, but we still have trouble treating sickle-cell disease.

The pressure is on, therefore, to translate the new science of molecular structure, of gen- and other 'omics, into clinical discoveries, into treatment of disease. What our leaders want is translation, NOW, and more translation TOMORROW. Sadly, as Borges quipped, "The original is always unfaithful to the translation."[14]

### Medea's Pharmakon

Translational research has been around longer than the NIH Roadmap. Indeed, if "bench-to-bedside" means conjuring up a useful drug to bring to the clinic, the Greeks were there first. They passed it on in the myth of Medea who brought colchicine, her *pharmakon* (φαρμακος) from a workshop of potions to the bedside of kings. Then as now, a pharmakon had the power to poison or to cure.

Medea was the daughter of Aeëtes, ruler of Colchis, a kingdom on the Black Sea in western Georgia. She presided over an Asian cult of potions extracted from herbs at the foothills of the Caucasus. In his *Medea*, Seneca describes the practice:

> *. . . Her hand harvests*
> *whatever earth creates in nesting spring*
> *or when brittle frost balds trees' beauty,*
> *forcing life inside itself with cold:*
> *grasses virulent with deadly flowers*
> *harmful juices squeezed from twisted roots.*[15]

Among the most potent products squeezed from those twisted roots was the juice of *Colchicum autumnale*, the yellow crocus of Colchis. Both historians of medicine and of botany suggest that the legendary "golden fleece" sought by Jason and the Argonauts was nothing but a mass of golden crocus, an Asian *pharmakon* needed in Europe to treat podagra, the gout of kings. Podagra, the swollen great toe of the gouty, was well described by Hippocrates and extracts of the golden crocus were known as both poison and cure.[16] On his quest to fetch the "golden fleece" from Colchis, Jason and the Argonauts set sail across the Aegean due east from Thessaly. They navigated the narrow straits of the Hellespont and Bosporos and braved the Black Sea in storm and tempest to land in Colchis. Once ashore, Jason was forced to perform a series of Herculean tasks set by King Aeëtes as price for the fleece. But Medea and Jason had become lovers and the princess used her potions to overcome the warriors and dragons that stood guard over the fleece. Jason returned to Greece bearing gifts: not only the golden fleece but also Medea, the sorceress who knew its powers as *pharmakon*.[17]

But Medea's charms were lost in translation. To the Greeks, Medea remained a foreign "healer" from Colchis, an outsider, a schemer, a MADEA (forgive us, Tyler Perry). Soon, however, Medea has translated her basic *pharmakon* into regenerative miracles and marital havoc at royal bedsides from Iolcus to Corinth. Setting an example for our own era, Jason deserted Medea to marry a younger princess, daughter of Creon ruler of Corinth. In a fit of murderous revenge, Medea poisoned not only the princess and Creon, but also slew her own two children in cold blood.[16]

The sorceress went unpunished—she was saved by solar energy. Medea's grandfather, the sun god Helios, sent a chariot powered by winged dragons to transport Medea and the bodies of her two children away to distant Athens. New amatory and pharmaceutical adventures awaited her; eventually her progeny founded a new land, Media, home to the Medes and Persians.

## Colchicine, Tubulin and the Japanese Iris

While the Greeks and Romans knew about the use of colchicine for gout and other disorders, the drug wasn't really available in pure form until the late 19th century, and problems with dosage, diagnoses and toxicities abounded. Gout was the stuff of legend and history, but colchicine remained a mystery: it stopped inflammation and the pain of acute gout but had no effect on tophi, those ugly deposits around the joints. Moreover, it didn't seem to work until the patient developed terrible diarrhea.[18]

But thanks to the pioneering clinical discoveries of Alfred Garrod, Dyce Duckworth and many others, a consensus was reached in the 19th century that colchicine was more or less specific for gout, that gout was caused by deposition of urate in joints and accumulation of these crystals resulted in tophi.[19] By 1889, Duckworth proposed that there was no more efficient agent than colchicines for acute gout and instituted the dosage regimen that has remained intact until today: treat to the point of diarrhea, and then cut treatment to minimum.[20] I might add that it was a century before a controlled clinical trial confirmed this method in 1987![21] It does make one wonder.

The next contribution, also in 1889, came from an Italian pathologist, who was looking for an agent that might reliably produce gastroenteritis. Suitably enough, it was a Sicilian achievement. B. Pernice found that when therapeutic doses of colchicine were given to experimental animals, lesions were produced in the nuclei of gastric and intestinal cells that had a remarkable appearance under the microscope: the cells were arrested in metaphase.[22] This translation to the pathologist's bench of a physician's observation (the first chap who wrote neatly of bedside colchicine was Alexander of Tralles in 580 CE) took over a millennium. More cause to wonder.

Things moved faster then, but not too fast. An American botanist and a Belgian pathologist, independently and then jointly, rediscovered the effects of colchicine on mitosis in plant and animal cells almost half a century after Pernice.[23,24] The colchicine explosion was on—in botany, pathology, oncology and finally in cell biology. By 1945, it was clear that the drug had major effects on the mitotic spindle, that it could produce metaphase arrest and polyploidy, the latter a boon in horticulture.[25] Indeed the Japanese irises in my garden, given to me by the late Currier McEwen, eminent rheumatologist and equally eminent iris fancier, owe their strength and deep color to tetraploidy.

Twenty postwar years later the biological revolution took up colchicine in those citadels of pretranslational research, the University of Chicago and the Marine Biological Laboratory at Woods Hole. In 1967, Ed Taylor and Gary Borisy used tritiated colchicine to identify the target of colchicine action in dividing and nondividing cells.[26,27] The protein they identified was the dimeric building block of microtubules, subsequently given the name "tubulin" by Mori.[28] The role of tubulins in excitable and nonexcitable tissue is now documented in over 18,500 publications in PubMed. We know now that the traffic of intracellular cargo in every cell in our body is carried on the tracks of microtubules. It's a two-way process, like translational research.[29]

When colchicine binds to tubulin, it stops the assembly of microtubules and then their integrity. But what about colchicine's major clinical use in gout, and its major toxic effects on the gut? Both are dependent on the interference by colchicine on a vital, microtubule-dependent process. The nature of that process? As someone who has tussled with the problem with respect to gouty inflammation, I must admit that although every possible sort of mediator from lysosomal hydrolases, to eicosanoids, to Toll-like receptor signals and "inflammasomes" have been implicated, we have suggestive clues but no real culprit.[30,31] And as for the gut? It seems that microtubules are needed to keep our intestinal cells pointed in the right direction, a function disrupted when some bacteria pretend they are colchicine.[32] But for a real explanation, we're still at sea and in dire need of new translations, or better yet, wisdom. One hopes that someone in one of those 60 translational research centers will find the real answer; whether by 'omics or luck, the way everything else in the story of colchicine and microtubules popped up. Perhaps Helios will come along with that chariot!

# Wiki-Science and Molière's Beast

*Many good men nowadays have been harmed by their desire to publish.*

—Molière, *The Misanthrope*, 1666[1]

*We wish to introduce WebmedCentral, a unique portal for rapid and free dissemination of biomedical knowledge through Post Publication Peer Review. It offers: Guaranteed publication of your research within 48 hours of submission. No pre publication peer review. Peer review takes place post publication in an open and transparent manner.*

Unsolicited e-mail, 2010[2]

*Is there some mechanism by which they can officially de-publish something that fails the "post-publication peer review?" I mean, if it stays published, then it's not peer review. Right?*

—James Sweet, 2010[3]

ELIMORE (Molière):
  *For fools contain inside of them a beast*
  *That triumphs when the world is made a fool!*

—David Hirson, *La Bête*, 1992[4]

## Wiki-Wiki Publication

2010 WAS A GOOD YEAR for the worldwide web of Wiki. Mark Zuckerberg of Facebook became *Time*'s Man of the Year, Julian Assange of WikiLeaks unleashed a flood of hidden documents and the *Los Angeles Times* explained to its remaining readers that "Science enables us to keep in contact with friends as well as release government secrets."[5] Well, I have a hunch that "science" was ahead of the *Times*. In September, many of us were solicited to submit original research and/or review manuscripts for a website called WebmedCentral.[2] Remind anyone of one of the NIH's PubMedCentral, or the popular WebMD?[6,7] Publication without review was guaranteed "within 48 hours," and peer review was guaranteed "in an open and transparent manner." The site clearly relieves authors of the painful process of responding to editors and reviewers; it also added a feature unique to scholarly publication:

> *Authors retain the copyright to the manuscript and are free to publish it elsewhere for a more targeted audience. . . . WebmedCentral does not have any problem with publication of previously published papers as the primary aim of this website is to facilitate scientific communication amongst biomedical researchers.*[8]

WebmedCentral promises new discoveries in biomedical science; and its venture into Wiki publication fulfills that promise. One finds on its site that smelling one's feet can prevent epileptiform seizures,[9] that vehicular accidents might induce fibromyalgia[10] and that beachgoers on Cancun have "a very high percentage of sunscreen use."[11] One can also learn about "Uner Tan syndrome" (quadripedal gait) from the evolutionary biologist who modestly named the syndrome by his own name: Uner Tan.[12]

Unlike other open web publications such as those of the PLoS group,[13] WebmedCentral follows unusual "post-publication review" practices. One finds corrections and comments directed at some, but not all, articles by self-selected reviewers—perhaps friends and family? At least one evanescent reviewer boasts that he has NO previous professional publications in the field reviewed.[14] No attempt has been made to judge the validity of the "reviews," to make calls on split decisions or to conclude that an original article is unprofessional; nor are doctoral degrees or academic positions required for either authors or reviewers.[14]

Some may judge these practices questionable, but WebmedCentral is fortunate in at least one contentious area. It is unlikely to be besieged by letters to the editor, since none is identified. This innovation would respect Elbert Hubbard's populist complaint that an editor's job "is to separate the wheat from the chaff and see to it that the chaff is printed."[15] Editor or not, carping reviewers or not, one thing is really for sure: on WebmedCentral your Wiki-paper is online in 48 hours: it may be unedited, but it's very, very quick! Wiki-Wiki one might say, since "Wiki" means quick, and "Wiki-Wiki" means *really* quick in Hawaiian, as Ward Cunningham explained when he named the Wiki-Wiki-Web after a Honolulu airport bus in 1995.[16]

## The Winner's Curse

I'll bet that other such efforts at unedited scientific publication will follow. In the twilight of printed journals and the zenith of Wiki-based text, it's a question of supply and demand. The traditional practice of manuscripts edited and reviewed by scholarly peers—who require revisions and corrections and who recommend acceptance or rejection to an identifiable editor—has come under attack.[17,18] Critics ask: if we can cull facts from a constantly changing Wikipedia, if we can trust our most private confessions to Facebook or Twitter, why *not* put all the news from the lab online? Why *not* open scientific publications to the wisdom of the marketplace? In these days of gene banks and genome dumps it's certainly possible to put every experimental result online in a sort of LabBookCentral. Why *not* let the reader—rather than editors or reviewers—decide whether what works and what doesn't at the worldwide bench?

The most thorough argument for such a sea change appeared in a PLoS article by Young et al. titled: "Why current publication practices may distort science."[18] They correctly describe the "extreme imbalance" between the availability of excess supply (the growing global output of biomedical science and clinical studies) and the limited demand for that output by a coterie of reputable scientific journals. The result is that only a small proportion of all research results are eventually chosen for publication, and these results do not truly represent the larger body of results obtained by scientists worldwide. They argue that

> the more extreme, spectacular results (the largest treatment effects, the strongest associations, or the most unusually novel and exciting biological stories) may be preferentially published. Journals serve as intermediaries and may suffer minimal immediate consequences for errors of over- or mis-estimation.[18]

This situation results in what economists who study auction behavior call the Winner's Curse. Academic auction theory holds that at any commodity auction, the winner necessarily pays more than a commodity is really worth: he overestimates demand and underestimates supply.[18] In scientific publishing, Young et al. imply that the few top journals tend to overestimate one or another piece of research and underestimate the global supply. Journals are more likely to publish work on a headline topic, by a

Nobel prize–winning author, from the richest lab over that of the unknown chap from Taiwan who has just discovered how zebrafish age. That which is published is "unrepresentative of scientists' repeated samplings of the real world."[18] Young et al. admit that the Winner's Curse theory is based on supply and demand behavior at oil auctions (post-Enron, pre-BP). They nevertheless suggest a number of radical remedies, among them the promotion of "rapid, digital publication of all articles that contain no flaws, irrespective of perceived 'importance'"[18] (i.e., Wiki-sites without adult supervision).

But after peer review of the material, an editor's job is precisely that of perceiving "importance." It's the job of copy editors—bless 'em—to make sure that articles "contain no flaws." Use of the term "Post-publication review" cheapens the word "review." As James Sweet noted, "if it stays published, then it's not peer review. Right?"[3] And if it hasn't been edited, it cheapens the word "edit." Right?

### La Bête and The JBC

I started worrying about the coming Wiki assault on language while watching the sparkling 2010 revival of David Hirson's 1992 play, La Bête (The Beast) on Broadway. Hirson, a contemporary American dramatist, wrote La Bête in rhyming verse, a novelty on the stage these days.[4] The play's hero is an elegant, forthright dramatist named Elimore, an anagram of Molière that dates to the mid-17th century, the period in which La Bête is set.[19] At a princely court, Elimore is suddenly confronted by the beast, Valere, a crowd-pleasing, self-referent, buffoon who wishes to join Elimore's theatrical troupe. As a sample of his "talent," Valere presents a confused, populist playlet for Elimore and the court to judge. It's wowed them in the sticks, he boasts. and he expects equal praise from the elite. But Elimore calls out the beast's little playlet for exactly what it is: a vacuous, rhetorical disaster written without an ear for language.

The central issue of La Bête is that of truth to the standards of one's craft: in written work, we call it editorial honesty. An analogy: Valere's proposal to Elimore/ Molière would be like a submission by a chiropractor now published by WebMedCentral[10] to the editors of The Journal of Biological Chemistry (JBC).

In pentameter, Elimore laces into Valere—who's played on Broadway as if he were a provincial auto-didact haranguing a Tea Party audience in Wasilla:

*A currency that's cheapened bit by bit*
*Til even genuine seems counterfeit*
*Hence, language is, by fools, emasculated!*
*Because their deeds and words are unrelated,*
*They taint all discourse with a hollow drivel:*
*Words, sapped of meaning, lose their clout, and shrivel!*
*Convictions, truths, are merely cloaks to don,*
*Providing an excuse to babble on!*[4]

Babble on or Babylon, whatever; the beast wins out. The royal patron rules that Elimore must share the troupe with the crowd-pleasing busker. But, ever true to his critical standards, Elimore declines, wandering off downstage into a solitary future as the play ends.

### The Misanthrope as Editor

Hirson, a Yale and Oxford–educated dramatist, has gotten his century and hero right. He's knowingly written a riff on Molière's *The Misanthrope*, a play in which truth to one's self is the central issue, as shown by the exercise of editorial judgment.[1] The misanthrope of the title is Alceste, who acts as the playwrights' alter ego.[19] In the jaded society of the day, he cannot find men or women who speak the truth openly, face problems directly or judge values fairly. He sees nothing but hypocrisy, deceit and flattery around him at court. Molière's elegant comedy weaves plot upon plot—of love won and lost and lost again—in a contest of rivals for the hand of sprightly Célimène. In the end, love's labor is forever lost in gossip and intrigue. Alceste retires from the field to solitude—as Elimore does in Hirson's retelling.

But the real battle is that of separating wheat from chaff. Molière confronts the question of how to tell a novice that the performance is not of concert grade. Oronte, a fellow noble—and fellow suitor for the damsel's hand—asks Alceste to read a sonnet Oronte has written. Alceste is bound by editorial honesty to tell Oronte to his face that the sonnet is a mess. He advises Oronte not to disgrace himself by publishing it—or anything else for that matter: "Hide these efforts from the public. . . . Frankly it would be better to leave it on your desk."[1]

Molière's Alceste was a skeptic at a time when skepticism and critical review of intellectual material was coming of age, and *The Misanthrope* was first performed in 1666, in the very year that the French Académie des sciences was launched by Colbert. It was written in the same year (1665) that the first modern scientific journals were founded—the *Journal des Sçavans* and the *Philosophical Transactions of the Royal Society* of London.[20] These journals set the stage for today's practice of scientific editing and review. In that tradition, Alceste's honest review of shoddy material should serve as a model for any modern skeptic confronting an unprofessional performance. He plays it as it lays, calling out junk for what it is, while praising to the skies that which is new and astonishing. He makes the perfect case for pre-publication review.

In the play, Alceste's brutal honesty costs him not only his friends but also the wife he sought. Célimène cannot bear to live outside the world of courtly flattery and gossip, while Alceste refuses to continue the trivial game of rank and fame at court. They split. At the end of the play, Alceste/Molière wanders offstage to a solitary life, to find—somehow, somewhere—a place to pursue excellence in *LUX et VERITAS*. Hinson has Elimore ring the the curtain down on his comedy:

> *If LIFE—not grim survival—is the aim,*
> *The only hope is setting out to find*
> *A form of moral discourse to reclaim*
> *The moral discourse fools have undermined.*[4]

# Arts and Science:
# Lewis Thomas and F. Scott Fitzgerald

*There's no evidence that Goethe ever had a "conflict" in the modern sense, or a man like Jung, for instance. You're not a romantic philosopher—you're a scientist, [Dick Diver] . . . You're bored with Zurich and you can't find time for writing here and you say that it's a confession of weakness for a scientist not to write.*

—F. Scott Fitzgerald, *Tender Is the Night*, 1934[1]

*I propose that classical Greek be restored as the centerpiece of undergraduate education. Latin should be put back as well, but not if it is handled, as it ought to be, by the secondary schools. English, history, the literature of at least two foreign languages, and philosophy should come near the top of the list, just below classics, as basic requirements [for medical school].*

—Lewis Thomas, 1979[2]

*By engaging for four years with the remarkable breadth of the liberal arts, students . . . have an opportunity to prepare themselves not for one profession but for any profession, including those not yet invented.*

—Shirley Tilghman, President of Princeton, 2010[3]

### Classics and Stuff

LEWIS THOMAS, OUR MOST ELEGANT ESSAYIST, wrapped up arts and science in one career. Witty, urbane and skeptical, Thomas may have been the only member of the National Academy of Sciences to have won both a National Book Award and an Albert Lasker Award. He is certainly the only medical school dean whose name survives on a professorship at Harvard, a prize at

Rockefeller University and a laboratory at Princeton—and whose books remain in print. Along the way, he worked out the science of immune surveillance, bacterial endotoxemia and intracellular proteolysis gone awry.[4]

Thomas's work owes much to his unique—and now quaint—coming of age at an Ivy liberal arts college in the early 1930s. His Princeton career followed a path trod by two fictional heroes of F. Scott Fitzgerald's, going from Amory Blaine, the Jazz Age rounder of *This Side of Paradise* (1920), to Dick Diver, the earnest neuropsychiatrist of *Tender Is the Night* (1934). Thomas and Fitzgerald can serve as models for the union of arts and science in a single, imaginative mind.

The Arts and the Sciences at one point did mean "Paintings and Stuff and Petri Dishes and Stuff," to quote Zadie Smith,[5] and for a good while, students at American colleges were expected to study both. But, just about the time Lewis Thomas suggested that we install Greek as a centerpiece of undergraduate studies, all that turned around.[6] Nowadays, educators rightly complain that "science and global competition have hollowed out the liberal arts."[7] In the last decade, undergraduate degrees in trade-oriented fields, such as "parks, recreation, leisure, and fitness studies," "security and protective services" and "transportation and materials moving," have been growing exponentially, all at the expense of general education.[8]

**Coming of Age at the Tiger**

In September of 1929, Lewis Thomas, not yet 16 years old, found himself sitting alone in a freshman dorm at Princeton. On his wire cot was a well-thumbed copy of Fitzgerald's *This Side of Paradise*, his schoolboy primer of college life. To any freshman of bookish bent, F. Scott Fitzgerald (Class of '17) was a literary idol, and the book's hero, Amory Blaine, became a guide to conduct among the *jeunesse dorée* of Princeton. Fitzgerald's book described how one went about attaining the grand precincts of Prospect Avenue with its smart clubs, easy booze and easier women. Edmund Wilson, the littérateur of the era, described his classmates returning to campus:

> *When I motored down to Princeton on May Day*
> *With John Bishop and Scott Fitzgerald . . .*
> *Poor Fitz went prancing into the Cottage Club*
> *With his gilt wreath and lyre*
> *Looking like a tarnished Apollo with the two black eyes*
> *That he had got, far gone with liquor, in some unintelligible*
>     *fight,*

> *But looking like Apollo just the same, with the sun in his*
> *yellow hair.*[9]

Fitzgerald's gilt wreath and lyre were notably absent on Dickinson Street, the mean quarters to which Lewis Thomas had been assigned. His digs were a reminder of how quickly Princeton separated its social sheep from unfit goats. Fitzgerald had it right on Amory's first day in a freshman rooming house:

> *"Oh, it isn't that I mind the glittering caste system," admitted Amory. "I like having a bunch of hot cats on top, but gosh, Kerry, I've got to be one of them."*
> *"But just now, Amory, you're only a sweaty bourgeois."*[10]

Lewis Thomas knew that Blaine would have regarded him as a sweaty bourgeois. He was, after all, a very young freshman, a doctor's son from Queens who had attended an obscure day school.[11] He shared quarters with ten other graduates of public or day schools, most of whom, like Lewis, had not been away from their parents for a single night. Most Princeton men were expected to have prepped at boarding schools, fewer to have attended private day schools and fewer still to have come from public schools. Not until 1956 did Princeton matriculate a class, the majority of which had attended public school.

For most of his freshman year, Thomas joined the fictional Amory Blaine in refusing to "get anywhere by working" at the required curriculum of Latin, math and English. As had Blaine, Lewis Thomas "turned into a moult of dullness and laziness, average or below in the courses requiring real work."[11] He began to take heavy interest in the several ways tobacco could be presented and dismissed athletics as a general waste of effort. However, high spirits and natural wit brought him inevitably to the offices of the *Princeton Tiger*, an impudent monthly that flourished in the heyday of college humor. Thomas knew that much of the witty stuff that ended up in *Vanity Fair* or *The New Yorker*, on the Broadway stage or Hollywood screen had roots in the *Princeton Tiger*, *The Yale Record*, the *Harvard Lampoon* and the *Jester of Columbia*. That was the league Thomas wanted to play in, and he soon found kindred spirits trying out for the *Tiger* board. They formed a cadre of clever wits.

At the *Tiger*, Thomas soon acquired a touch of Coward (Noël), a pinch of Porter (Cole) and a dose of Held (John Jr.). His particular mentor was the future *New Yorker* cartoonist, Whitney Darrow Jr. (Class of '31), a fine-tuned humorist whose undergraduate performance was already of concert grade. Thomas flourished under Darrow in the *Tiger* milieu, and by the end of his first year, he knew that he'd come a long way from that lonely cot in the freshman dorm. He also knew that he'd better study the classics hard to keep up with the "toffs" who'd gotten their Horace in prep school.

## Darwinism on Prospect Avenue

Princetonians ate in common dining rooms until their fourth semester, but by sophomore year, they were faced with "bicker," a rite of initiation into the various Prospect Avenue clubs. As the prestige of one's club determined one's place in the social scheme, an anxious sophomore would sit in his room at bicker time, biting his fingernails, anxiously awaiting the moment when the bicker committee of one or another club would knock on the door to vet his manners, wardrobe and genealogy.

Thomas was passed over by the grander clubs and grateful to be picked for the Key and Seal Club, although he knew he had been picked for a club that was, literally, the furthest out on Prospect Avenue. Nevertheless, it was *on* Prospect Avenue, a tree-clad boulevard of dreams that came to life on football weekends and P-rades, which featured signs proclaiming, "Gentlemen Prefer Bonds—Others Sell Insurance."[9]

The stock market crashed barely one month after Thomas entered Princeton, but club life on Prospect Avenue went on without skipping a beat. Banks may have failed in Yonkers, but to paraphrase Ira Gershwin, Princetonians were sure that posterity was just around the corner. The horseplay and alcohol were a constant source of friction between the young bloods and the dean of the college, Christian Gauss, a scholar much admired by young Thomas. But on Prospect Avenue, one soon learned that if you wanted to fit in, you shut up about literature and drank your booze. After the crash of 1929, the boys at the clubs drank bootleg gin à la Scott and wife, Zelda, who "went out like a light" after jumping into one or another of New York's fountains to launch the Jazz Age. Thomas did a piece for the *Tiger* about that—"Out Like a Light."[12]

Thomas's classmates, like Fitzgerald's, were convinced that without clubs, there would be no Princeton. That conviction trumped wiser counsel. When president of Princeton, in 1907, Woodrow Wilson had called the life of those excluded by the clubs "a little less than deplorable," warning that Princeton itself was in danger of becoming simply a "background for life on Prospect Avenue."[13] Wilson's plan to dismantle the system was defeated by powerful alumni who added a twist of social Darwinism to simple class war-fare. Henry Fairfield Osborne (Class of 1877, eugenicist and a great fan of the Master Race) spoke for the opponents of Wilson when he argued that:

> *Competition is not confined to human rivalries and struggles;*
> *it pervades the whole animal kingdom of life; it is the basis of*
> *Darwin's doctrine of evolution; it has been, and ever will be,*
> *the means of progressive evolution. . . .To extinguish the spirit*
> *of competition is to seek racial suicide.*[13]

It took two wars, but Gauss and the liberal scholars prevailed. Social Darwinism may be alive and well in other parts of our country, but Princeton today is a diverse campus, where distinction in the classics has replaced class distinction. There's also that building named for Lew Thomas, the "sweaty bourgeois" who came of age in the presence of both art and science.

## Arts to Science

Like many of his classmates, Thomas distrusted untested thought and political rhetoric. He acquired a permanent disdain for dense Germanic thought, recalling that "the only sure memory I retain of Heidegger . . . is pure bewilderment." On the other hand, he lost his heart and soul to lyric poetry, gobbling up chunks of Longfellow, Holmes, Whitman and the English Romantics. "I once knew Keats, lots of Keats by heart . . ."[14] as in *Ode to a Nightingale*:

> *But on the viewless wings of Poesy,*
> *Though the dull brain perplexes and retards:*
> *Already with thee! tender is the night,*
> *And haply the Queen-Moon is on her throne.*[15]

Tender is the night, indeed. He also toyed for hours with his own attempts at humorous verse and devoured the early work of Evelyn Waugh.

In his senior year, a new passion supervened; Lewis Thomas finally took wings as a "reasonably alert scholar."[11] He attributed this change to an advanced biology course with Professor Wilbur Swingle, an eminent physiologist who taught Thomas two lessons that stayed with him for a lifetime. The first of Swingle's teachings was that science begins with the admission of ignorance. Thomas often repeated that admonition, arguing that it would be useful to teach courses in medical ignorance, a curriculum he promoted as a

> *new and subversive technique for catching the attention of*
> *students driven by curiosity, delighted and surprised to learn*

*that science is . . . an "endless frontier." The humanists, for
their part, might take considerable satisfaction watching their
scientific colleagues confess openly to not knowing everything
about everything. And the poets, on whose shoulders the future
rests, might, late nights, thinking things over, begin to see some
meanings that elude the rest of us.*[16]

The second of Swingle's lessons was more direct. Thomas learned that
experiments done for the sake of curiosity often yield the most practical
results. Swingle's work showed the way. The question of how endocrine
glands prepare an organism for survival was a hot item in the early 1930s,
a puzzle ready to be solved. Walter B. Cannon had shown that adrenalin
of the adrenal medulla played a key role in the "fight or flight" response.
But what about that other part of the gland—the adrenal cortex? In 1929,
Wilbur Swingle and a student, J. J. Pfiffner, published a paper in *The Anatomical Record* showing that an aqueous extract of the cortex could keep
dogs alive after their adrenals had been removed.[17] It was the beginning
of hormone replacement therapy in Addison's disease, caused in those days
chiefly by tuberculosis. Swingle and Pfiffner rapidly followed up this paper
with another in *Science*.[18] Swingle's extract normalized the body's delicate
balance between sodium and potassium ions, a ratio shown by Jacques Loeb
to be crucial for life.[19] Swingle's work pointed the way to the isolation of
deoxycorticosterone (DOC) and eventually, to the even bigger discovery of
cortisone. It had a major influence on Thomas's later experiments with cortisone and inflammation, for which Lewis Thomas received international
recognition.[20]

## Two Masterpieces

Perhaps for these reasons or perhaps simply because he was growing up
and becoming wiser, Lewis Thomas became alert and attentive and began to
think seriously about science in general and biology in particular. He once
said that he discovered in Swingle's classroom that biology was the science
that made it intellectually respectable to become a doctor.[11] By the end of
his senior year, Thomas had become a Biology Watcher.

Impatient with the "romantic philosophies" he'd been taught by humanists at the college, Thomas became a convert to the reductionist views
of Edwin Grant Conklin (1863–1952), chairman of the Princeton biology
department and fierce opponent of Osborne's overt racism. A student of in-
vertebrate embryology at Princeton and at the Marine Biological Laboratory

at Woods Hole, Conklin not only was a Woods Hole colleague of Jacques Loeb but also shared his liberal mechanistic philosophy. Conklin, like Loeb, investigated the physical chemistry of cell division and became an authority about human evolution, popularizing his science in books such as *The Direction of Human Evolution* (1921) and *Problems of Organic Adaptation* (1921). Conklin's irreverent disdain for the notion of a "designed universe" caught Thomas's fancy. One of Thomas's undergraduate compositions, which appeared in October of his senior year, is titled "A Disrespectful Note on the Divine Plan." "L.T." (Thomas) turns his back on theology and informs us that

> *By sinning we get lots of things*
> *To entertain us when we're bored,*
> *But labored virtue only brings*
> *Virtue as its own reward*
> *I think it futile of the Lord*
> *To say that goodness should result*
> *In such a slim reward,*
> *When goodness is so difficult.*[21]

Goodness was difficult indeed for Scott and Zelda. The Jazz Age had crashed for both of them: alcoholism and schizophrenia disrupted their golden lives. In the year that Thomas was discovering Conklin and Loeb at Princeton, Fitzgerald had learned the darker lessons of mental disease from Freud and Jung. Yet, from the tortured years with poor Zelda, Fitzgerald forged his masterpiece, *Tender Is the Night*. Its hero is not a Jazz Age rounder nor a romantic philosopher, but the physician/scientist Dick Diver.

There is a footnote to Lew Thomas's conversion of Amory Blaine to Dick Diver. In the early 1970s, Thomas spent his summers at Woods Hole, where he studied the effects of endotoxin on amoebocytes of the horseshoe crab (*Limulus*). The house Thomas bought at Woods Hole was only a few blocks removed from the old Conklin house on High Street, and the building in which he punctured crabs was the Loeb building. He had also come to the seashore to collect his thoughts and his essays. *His* masterpiece, *The Lives of a Cell*, written in good part at Woods Hole, appeared in 1974. The rest is part of the history of arts and science.

# Icarus and Fukushima Daiichi: Human Factors in a Meltdown (Sv=1J/kg.w)

BODY RADIATION TESTED. *Dr. Rolf Sievert of the radiophysics department of the Caroline Institute in Stockholm, who constructed the apparatus, says the exact measurement of radiation represents the first step in its ultimate control.*

—*The New York Times,* April 5, 1952[1]

*Prime Minister Naoto Kan said his government was in a state of maximum alert over the situation at the Fukushima Daiichi Nuclear Power Station . . . A leak of the radioactive water, which measured more than 1,000 millisieverts (mSv) an hour on the surface at Reactor No. 3, could exacerbate smaller leaks of radiation detected in the seawater around the plant.*

—*The New York Times,* March 30, 2011[2]

*The thermal Springs of Ikaria [named for Icarus] are considered as some of the most radioactive springs in the whole world [and] the most suitable natural way for healing several diseases; some of these are: Chronic rheumatisms and several forms of arthritis, Uric arthritis, Neuralgia, neuritis, and muscle pains, etc. etc.*

—Isle of Ikaria Holiday Brochure 2011[3]

*The working personnel in the thermal spa installations are exposed to significant radiological risk due to waterborne $^{222}Rn$ with a maximum dose rate up to 35 millisieverts $y^{-1}$, which led to overexposure in terms of the 20 millisieverts Sv $y^{-1}$ professional limits.*

—Report on "therapeutic spas" in Ikaria, Greece, 2010[4]

## Sunami, Shock and Disbelief

THE DISASTER THAT HIT JAPAN ON MARCH 11, 2011, was unique in the atomic age: an earthquake, followed by a tsunami, followed by a meltdown. There have been other earthquakes, other tsunamis and other nuclear accidents, such as Chernobyl, but the calamitous sequence at Fukushima Daiichi had not happened before. It put full stop to arguments both for and against nuclear energy; time had come to cope with the crisis, to monitor the site and to gauge its effects on humans and the environment. A clear, rapid account was expected so that "Civil society in all countries could assess the record of nuclear power and draw the conclusions."[5] It didn't quite play out that way.

Japanese authorities followed the playbook of other national disasters (the BP oil spill in the Gulf or Hurricane Katrina). They went through the stages of administrative grief: Shock and Disbelief followed by Denial and Anger, Bargaining and Guilt. They issued frequent and often contradictory accounts, but pretty soon the facts of the meltdown emerged. The logs of the International Atomic Energy Agency show that the world had been watching and counting.[6]

On March 11, Japanese authorities claimed that the disaster had caused "no release of radiation from any of the nuclear power plants affected by today's earthquake and aftershocks." And on March 15, they insisted "there has so far been no release of radiation from any of the nuclear power plants." But massive douches of seawater had failed to back up the crippled cooling systems in all functioning reactors. Hydrogen explosion followed hydrogen explosion, fires broke out to release $I^{131}$ and Cesium$^{137}$ into the air and seas. By March 22, it became clear what had really happened inside the plant: three of the six reactors were in advanced stages of meltdown. The zirconium alloy sheaths that surround the reactor's fuel rods had ruptured, and pellets of molten uranium dropped to the bottom of the reactors, where they congealed. The core of Unit 4 landed in a cooling pond in open air after two hydrogen explosions macerated the building.[6,7] All that water added to cool the cores had overheated and needed extrusion. The pot was boiling.

## Becquerels and Sieverts

One month after the tsunami hit, the bad news was spelled out in becquerels and sieverts. The total radioactivity released was between 360,000 and 600,000 trillion becquerels, and at least 21 workers were exposed to over 100 millisieverts.[8]

These units require explanation: a becquerel is 1 disintegration of a radioactive substance per second and describes the total amount of radioactive energy released. A sievert, on the other hand, expresses the effect of radiation on humans. A sievert (Sv) is the energy (J) absorbed by an amount (kg) of human tissue exposed to various sources of radiation (w). Therefore 1Sv=1J/kg.w and that "w" is the human factor. Humans are exposed to natural background radiation of 2.4 mSv per year; we tot up an extra 3 mSv during a mammogram or 5 to 30 mSv from a CT scan. In most countries the current maximum permissible dose to radiation workers is 20 mSv per year averaged over five years, with a maximum of 50 mSv in any one year.[9]

That's why observers were concerned when on April 4 a single crack in the pit at the No. 2 reactor generated readings of 1,000 millisieverts (1 sievert) of radiation per hour in the air. Forced to release the cooling water from the pit into the sea, engineers found it to contain one million becquerels of $I^{131}$ per liter—10,000 times normal in cooling water at a nuclear plant. Meltdown! The numbers got worse. On April 12, the government announced it had raised the severity of the Fukushima-Daiichi incident from level 5 (Three Mile Island level) to level 7, the worst (Chernobyl level) on the international scale. One official continued to deny the worst:

> *Some foreigners fled the country even when there appeared to be little risk, if we immediately decided to label the situation as Level 7, we could have triggered a panicked reaction.*[10]

Finally, the Japanese authorities admitted to a total escape of 370 trillion becquerels, but "only 20 percent" of the total emitted at Chernobyl. Independent experts came up with closer to 630 trillion becquerels. Quibbling ensued, final estimates varied and the authorities raised the maximum permissible exposure of Japanese nuclear plant employees to 250 mSv/year.[11] Meanwhile, the seas and marine life around the plant were bathed by five trillion becquerels.[12]

The long-range effects of this accident cannot be projected. $I^{131}$ decays within eight days but $Cs^{137}$ has a half-life of 30 years and is likely to be with us for a good while. We simply don't know enough about the biology of low-dose radiation exposure over time, over populations and floating in the seas. Indeed, the only research program that supports the basic science behind low-dose radiation risks, at the U.S. Department of Energy, is facing heavy cuts, or even elimination.[13] As Columbia's David Brenner has pointed out:

> *Soon we will either have to replace many of our aging fleet of*
> *reactors or move away from nuclear power entirely. To make*
> *rational decisions about these questions—and what we should*
> *do about the rapid increase in medical imaging such as CT*
> *scans, or even the new airport X-ray scanners—we need to*
> *understand the risks of low doses of radiation with a great deal*
> *more certainty.*[13]

### Sievert of the Karolinska

What we do know about the effects of radiation on humans we owe in large part to the work of Rolf Maximilian Sievert, who devised the instruments by which we measure radiation doses and who found ways to protect doctors and patients from the ill effects of X-rays and radium. He figured out that humans and their tissues pose a complex target for radiation stress, a story that illustrates the undergraduate aphorism:

> *Physicists define stress as force per unit area, the rest of us de-*
> *fine stress as physics.*

Sievert was born in Stockholm in 1896. He studied at the Karolinska Institutet, the Royal Institute of Technology in Stockholm, and earned his graduate degree from Uppsala.[14] His 1932 masterwork, "A Method for Measuring Röntgen, Radium and UV rays and Some Experiments on the Utility of this Method in Physics in Medicine,"[15] earned him a PhD, great praise in the *British Medical Journal*[16] and a professorship at Karolinska Institutet. In 1924, he founded the Radiumhemmet (Home for Radium), which became an international center for radiation physics and cancer treatment. By 1938 the Hemmet became the Department of Radiation Physics at Karolinska Hospital, which Sievert developed and which thrives today.[14]

His major invention was the Sievert chamber—or chambers—a series of ever evolving devices for measuring ionizing radiation. The original chamber, developed in the late 1920s, was a pair of concentric cylinders of a magnesium alloy with an electrode attached. When such a chamber—with a preset electronic charge—is exposed to ionizing radiation, ion pairs form in the air between the two cylinders and the reduction in charge is a measure of the radiation received by the chamber.[14] Sievert chambers, with their many modifications—the latest in the 1950s—have been used to measure the absorbed energy of radiation in body cavities, fluids, tissues and entire organs. Their use has been extended to monitor radium and radiology suites the world over, to measure radiation in nuclear submarines, supersonic jets and travel in space. They were used in the robots that threaded their way into the reactors at Fukushima. Sievert's main contribution however, was not a device but the proposition that humans are living material made of stuff like blood and bone that differ in radiosensitivity.

Before Sievert, one knew that radioactivity exerted a force, expressed in joules (J), directly related to disintegrations per second in becquerels (Bq). But to determine the total energy absorbed by an actual object, such as a plate of glass, or a patient, the weight of that object (k) had to be factored in. That unit 1J/kg—known as a gray—is OK for deciding what was absorbed by object of any size, but humans differ in two ways. Living beings differ in their response to beta or gamma rays, and different tissues differ in their response to the same source of radiation: skin is less sensitive than are gonads, bone differs from blood, etc. Each of these factors derived experimentally at the Hemmet and elsewhere is regularly integrated into the weighing factor "w"—the human factor. In 1979, the sievert was internationally recognized as the "equivalent and efficient dose" of radiation absorbed by humans: Sv=1J/kg.w.[17]

### Radon$^{222}$ for Fibromyalgia

In the spring of 2011, while thermal communities along the Japanese shore worried whether they were zapped by the five trillion becquerels of radioactivity spilled in their waters, other holiday communities touted their radioactive waters as just fine for tourists with sore joints. Italy's Montecatini Terme promised that "wrapped in mud packs, drinking mineral-laden waters, getting radioactive vapors steamed into your face you will forget all about the stress."[18] Not to be outdone, the craggy Isle of Ischia in the Bay of Naples claims that its "radioactive thermal baths, are reputed to help ailments such as arthritis [but] it takes a local doctor's prescription to find out."[19] In Ischia's thermal baths the active ingredients are Rn$^{222}$ and Rn$^{220}$, which come down the disintegration chain from uranium and radium.[20]

Austria is in the picture too: in Bad Gastein you can treat your fibromyalgia with Radon$^{222}$ either in its gaseous or liquid form—depending on whether you wish to ride your "w" on trolleys into radioactive caverns (Radon content of 44 Bq/l of air) or simply take the baths.[21] In the waters of Bad Gastein, radiobiologists found that:

> Doses ranged from 0.12 mSv in the thermal bath to 0.33 mSv
> in the vapour bath, exceeding equivalent doses to the inner

*organs (kidneys) by inhaled radon and progeny by about a factor 3, except for the lung, which receives the highest doses via inhalation.* [22]

None of the websites touting radioactive healing have been affected by the meltdown at Fukushima. They may not cure fibromyalgia or "arthroses" but they may well have effects we know nothing about. David Brenner is right: we do need to understand the risks of low doses of radiation with a great deal more certainty.

### Icarus Melting Down

Most disturbing is the claim for isotopic healing made by Ikaria, whose tourist guide boasts, "Along Ikaria's coastline there are many areas where radioenergic hot mineral springs flow into the sea from the shoreline where it is possible for one to bath/swim." [23] Actually, while a tourist might not really suffer more absorbed radiation from those bath/swim sessions than from a chest X-ray, the poor folk who work at the thermal spa installation are exposed to significant radiological risk due to waterborne $Rn^{222}$ with a maximum dose rate up to 35 mSv/y. [4] The Greek island was named for the dead Icarus of legend who washed up on its shores after solar radiation induced his meltdown. Daedalus, the great inventor, had fashioned wings of feathers and wax and taught his son Icarus to fly. Urged ever upward by his father, Icarus flew too close to the sun; the wax that held his wings together softened in solar heat and he plummeted into the Ikarian sea. [24] It's a legend to remember anytime we play with fire, without quite knowing all we need about thermodynamics.

That's why we need the Sieverts of this world to point out that "the exact measurement of radiation represents the first step in its ultimate control."

# Acknowledgments

THE ESSAYS IN THIS VOLUME have been modified from the essay/editorials I publish monthly in *The FASEB Journal*, the official journal of the Federation of American Societies of Experimental Biology, a publication for which I am responsible. I acknowledge the terrific support given me by the FASEB Office of Publications with Jennifer Pesanelli, its present director; Cody Mooneyhan, associate director for journal services and managing editor of *The FASEB Journal*; together with Mary Hayden, Susan Moore and Gail Fallon. That office has been home to the most diligent, literate unit ever to grace the field of scientific communication.

As in most of my published work, Ms. Andrea Cody, the administrator of the Biotechnology Study Center at the NYU School of Medicine, has kept the flame of critical judgement alive, and made it possible to turn a collection of essays into a coherent volume.

Finally, I am always amazed how the Bellevue Literary Press manages to work at the Pulitzer Prize level (2010) with modest resources in an unusual setting. This feat of editorial magic has been accomplished by the astounding Erika Goldman, who has overseen several of my books and whose tireless dedication to the written word is worth a million tweets.

# References

**Epigraphs**

1. Forster, E. M. (1911, 1997) *The Machine Stops and Other Stories* (London: André Deutsch), p. 87ff

2. Stengel, R. (2010) *Time's* Man of the Year: Only Connect (essay) *Time,* Dec. 15, p. 23

**Prefatory Note**

1. Smith, Z. (2003) *The Autograph Man* (London: Penguin), p. 19

2. http://www.charlierose.com/guest/view/210; http://twitter.com/charlierosetwitter.com/jayz; http://www.charlierose.com/guest/view/881

3. Jaenisch, R., Bird, A. (2003) Epigenetic Regulation Of Gene Expression: How the Genome Integrates Intrinsic and Environmental Signals, *Nature Genetics* **33**, 245–254

4. Changeux J-P., Ricoeur, P. (2002) *What Makes Us Think: A Neuroscientist and a Philosopher Argue about Ethics, Human Nature, and the Brain,* tr. M. B. DeBevoise (Princeton: Princeton University Press), p. 239

5. Hammett, D., Hackett, A., Goodrich, F. (1939) *Another Thin Man,* MGM film

**Walter Benjamin and Biz Stone:**
**The Scientific Paper in the Age of Twitter**

1. Dowd, M. (2009) To Tweet or Not to Tweet. *New York Times* http://www.nytimes.com/2009/04/22/opinion/22dowd.html. Accessed May 2009

2. Benjamin, W. (1935) The Work of Art in the Age of Mechanical Reproduction. In *Illuminations.* Hannah Arendt, (New York: Schocken, 1969) p. 232. The essay was first published in French—as "L'Oeuvre d'Art à l'Époque de sa Reproductibilité Technique"—by Max Horkheimer in the Institute for Social Research's journal, *Zeitschrift für Sozialforschung*

3. Zivkovic, B., DeRisi1, S., Real, B., Brunton, T., Cave, R., Uman, R., Toomey, S., Patterson, M., Suttor, J., Kirton, J., Worden, A., Kapoor, A., Krishnan, P., Tschalär, R. (2006) PLoS Journals Sandbox: A Place to Learn and Play. *PLoS ONE* **1**(1),e0doi:10.1371/journal.pone.0000000. Accessed May 2009 [CrossRef]

4. Lenhart, A. (2009) Twitterpated: Mobile Americans Increasingly Take to Tweeting. *Pew Internet & American Life Project* http://pewresearch.org/pubs/1117/twitter-tweet-users-demographics. Accessed May 2009

5. Noonan, E. (2009) Life Is Tweet. *Boston Globe* http://www.boston.com/lifestyle/articles/2009/01/04/life_is_tweet/. Accessed May 2009

6. Lindsay, J. (2009) Boston Train Accident Prompts Strict Ban on Operators' Cell Phones, Move Comes After Driver of Crashed Trolley Admits He'd Been Texting Girlfriend. *Houston Chronicle*, May 10, 12

7. Miga, A. (2009) John Kerry: Newspapers "Endangered Species." Associated Press http://www.huffingtonpost.com/2009/05/06/john-kerry-newspapers-end_n_197869.html. Accessed May 2009

8. Bar-Ilana, J., Finka, N. (2005) Preference for Electronic Format of Scientific Journals—A Case Study of the Science Library Users at the Hebrew University. *Library and Info. Sci. Res.* **27**, 363–376

9. Committee on Electronic Scientific, Technical, and Medical Journal Publishing, The National Academies (2004) *Electronic Scientific, Technical, and Medical Journal Publishing and Its Implications: Report of a Symposium* http://www.nap.edu/catalog.php?record_id= 10969. Accessed May 2009

10. Kronick, D. A. (1976) *A History of Scientific & Technical Periodicals* (2d ed.: Metuchen, NJ: Scarecrow Press)

11. Darwin, C., Wallace, A. (1858) On the Tendency of Species to form Varieties; and on the Perpetuation of Varieties and Species by Natural Means of Selection. *Proc. Linnaean Soc.* **3**, 45–62

12. Southgate, D. A. T. (1995) The Structure of Scientific Papers. *Br. J. Nutr.* **74**, 605–606 [CrossRef]

13. Benjamin, W. (1935) The Work of Art in the Age of Mechanical Reproduction, p. 232

14. Scher, S. (2004) Was Watson and Crick's Model Truly Self-Evident? *Nature* **427**, 584 [Medline]

15. Watson, J. C., Crick, F. H. C. (1953) A Structure for Deoxyribose Nucleic Acid. *Nature* **171**, 737–738 [CrossRef] [Medline]

16. Benjamin, W. (1989) *The Arcades Project*, H. Eiland and K. McLaughlin (Cambridge, MA: Harvard University Press), p. 1007

17. Valéry, P. (1928) La conquête de l'ubiquité. (1960) *Oeuvres, tome II, Pièces sur l'art*, NRF (Paris: Gallimard, Bibl. de la Pléiade), p. 3

18. Rollason, C. (2002) *Border Crossing, Resting Place: Portbou and Walter Benjamin* http://www.wbenjamin.org/portbou.html. Accessed May 2009

**Epigenetics in the Adirondacks**

1. Darwin, C. (1859) *On the Origin of Species by Means of Natural Selection, or the Preservation of Favoured Races in the Struggle for Life* (London: John Murray)

2. Agassiz, L. (1869) *De l'Espèce et de la classification en zoologie* (Paris: Ballière), pp. 375–391

3. Woolf, B. (1955) On C. H. Waddington's 50th Birthday. In Robertson, A. (1977) Conrad Hal Waddington. *Biogr. Mem. Fellows Royal Soc.* **23**, 575–622

4. Fell, H. B. (1960) Fashions in Cell Biology (Presidential Address International Congress Cell Biology, Paris). *Science* **132**, 1625–1627[Free Full Text]

5. Goldberg, A. D., Allis, C. D., Bernstein, E. (2007) Epigenetics: A Landscape Takes Shape. *Cell* **128**, 635–638 [CrossRef] [Medline]

6. Youngson, N. A., Whitelaw, E. (2008) Transgenerational Epigenetic Effects. *Annu. Rev. Genomics Hum. Genet.* **9**, 233–225 [CrossRef]

7. Mungall, A. J. (2002) Meeting Review: Epigenetics in Development and Disease. *Comp. Funct. Genomics* **3**, 277–278 [CrossRef] [Medline]

8. Leake, J. (2008) How Your Behaviour Can Change Your Children's DNA. *Sunday Times (London)*, July 20, 16

9. Leperchey, F., Barbet, J. P. (1998) The Origins of Embryology. *Morphologie* **82**, 19–28[Medline]

10. Mayr, E. (1982) *The Growth of Biological Thought: Diversity, Evolution, and Inheritance* (Cambridge: Harvard University Press)

11. Dawkins, R. (1976) *The Selfish Gene* (Oxford: Oxford University Press)

12. Wilson, E. O. (2008) One Giant Leap: How Insects Achieved Altruism and Colonial Life. *Life BioScience* **58**, 17–25

13. Guil, S., Esteller, M. (2008) DNA Methylomes, Histone Codes and miRNAs: Tying It All Together. *Int. J. Biochem. Cell Biol.* [Epub ahead of print] (2009) *Int. J. Biochem. Cell Biol.* **41** (1), 87–95 [Epub 2008 Sept 13]

14. Vincent, A., Ducourouble, M.-P., Van Seuningen, I. (2008) Epigenetic Regulation of the Human Mucin Gene MUC4 in Epithelial Cancer Cell Lines Involves Both DNA Methylation and Histone Modifications Mediated by DNA Methyltransferases and Histone Deacetylases. *FASEB J.* **22**, 3035–3045[Abstract/Free Full Text]

15. Lederberg, L. (2001) The Meaning of Epigenetics. *The Scientist* **15**, 6

16. Ptashne, M. (2007) On the Use of the Word "Epigenetic." *Current Biology* **17**, R233–R236 [CrossRef] [Medline]

17. Waddington, C. H. (1939) *An Introduction to Modern Genetics* (London: Allen & Unwin), pp. 154–156

18. Robertson, A. (1977) Conrad Hal Waddington. *Biogr. Mem. Fellows Royal Soc.* **23**, 575–622[CrossRef]

19. Waddington, C. H. (1942) The Epigenotype. *Endeavour* **1**, 18–20[CrossRef]

20. Jamieson, P. (1982) *The Adirondack Reader* (Glen Falls, NY: Adirondack Mountain Club)

21. Emerson, R. W. (1918) The Adirondacs [*sic*]. *The Complete Works of Ralph Waldo Emerson* (Boston: Houghton Mifflin), vol. 3, p. 296

22. Schneider, P. (1996) *The Adirondacks*, (New York: Henry Holt), p. 191

23. Stillman, W. J. (1901) The Philosophers' Camp 1858. *Autobiography of a Journalist* (Boston: Houghton Mifflin), vol. 1, p. 271

24. Agassiz, E. C. (1886) *Louis Agassiz: His Life and Correspondence* (Boston: Houghton Mifflin), p. 132

## A Nobel Is Out of Order: "J.Lo" vs. Hypathia of Alexandria

1. Stein, R. (2010) Robert Edwards Wins 2010 Nobel Prize in Medicine for In-Vitro Fertizilation. *The Washington Post,* Oct 4, p. 1

2. Smith, R. (2010) Father of IVF Who Gave Hope to the Childless Wins Nobel; Vatican Calls Choice "Out of Order." *Daily Telegraph* London, Oct. 5, p. 1

3. Rainey, C. (2010) Jennnifer Lopez. *Elle,* Jan. 5 http://www.elle.com/Pop-Culture/Cover-Shoots/Jennifer-Lopez/Jennifer-Lopez-ELLE-Magazine-Interview-and-Photos-February-2010

4. Draper, J. W. (1864) *A History of the Intellectual Development of Europe* (New York: Harper Bros.), p. 228

5. Starr, M. (2010) Out of Order; JLo, Tyler: Simon Make It Look Easy. *New York Post,* Sep 30, p. 81

6. Irwin, N. (2010) Nobel Economics Prize: Peter Diamond, Dale Mortensen, Christopher Pissarides Share Award." *The Washington Post,* Oct. 11, p. 1

7. Branigan, T. (2010) Liu Xiaobo Nobel Win Prompts Chinese Fury. *Guardian,* Oct. 9, p. 4

8. Anonymous (2010) Chinese Agency Blasts Liu Xiaobo's Nobel Award. *BBC Monitoring Asia Pacific,* Oct. 16

9. *Anonymous* (2010) Vatican Health Experts "Dismayed" by Nobel Prize for IVF Co-Developer, Oct. 5 *www.catholicnewsagency.com/.../vatican-health-experts-dismayed-by-nobel-prize-for-ivf-co-developer/*

10. Edelson. A. M. (2006) Nobel "no-shows." *FASEB J.* 2006 **20,** 3–6

11. Jha, A (2010) A British IVF Pioneer Robert Edwards Wins Nobel prize for Medicine, *Guardian,* Nov 4, p. 1 http://www.guardian.co.uk/science/2010/oct/04/ivf-pioneer-robert-edwards-nobel-prize-medicine

12. Kois, D. (2010) Aww, C'mon, J-Lo Deserves a Little Love. *The Washington Post,* Apr. 23

13. Goldstein, J. (2001) Presentation of 2001 Lasker Award http://www.laskerfoundation.org/awards/2001_c_presentation.htm

14. Kass, L. R. (1971) Babies by Means of In Vitro Fertilization: Unethical Experiments on the Unborn? *N Engl J Med.* **285,** 1174–79

15. Andrews, L. B. (1999) *The Clone Age* (New York: Henry Holt), p. 12

16. Heneghan, T. (2010) IVF Discovery Opened Pandora's Box of Ethical Issues. *Reuters,* Oct. 4

17. Anonymous AP (1989) 4 Religions Put Limits on Test Tube Babies: Catholics, Jews, Muslims and Buddhists Oppose Human Embryo Experiments. *Toronto Star,* Apr. 6, p. A23

18. George, R. (1979) "Test Tube Babies": Exploring Ethical Question, *Jewish Press,* May 11, p. 7

19. Pope Paul VI (1968) *Humanae Vitae* http://www.papalencyclicals.net/Paul06/p6humana.htm

20. Franklin, J. L. (1987) Vatican Raps New Methods of Procreation. *Boston Globe,* Mar. 10, p. 1

21. Hooper, J. (2008) Vatican Condemns IVF in Bio-Ethics Review. *Guardian,* Dec. 13, p. 25

22. Pope Pius XII (1950) *Humani Generis* http://www.papalencyclicals.net/Pius12/P12HUMAN.HTM

23. Pope Leo XIII (1895) *Adiutricem* http://www.papalencyclicals.net/Leo13/l13adiut.htm

24. Chapman, J. Cyril of Alexandria. In *The Original Catholic Encyclopedia* (online edition) http://oce.catholic.com/index.php?title=Cyril_of_Alexandria,_Saint

25. Amira, D. (2010) GOP's Delaware Senate Nominee Christine O'Donnell Not a Big Fan of Evolution. *New York Magazine* (online edition) Sept. 15 http://nymag.com/daily/intel/2010/09/the_gops_delaware_senate_nomin.html

26. Dzielska, M. (1996) *Hypatia of Alexandria* trans, F. Lyra (Cambridge, MA: Harvard University Press), p. 2

27. Hubbard, E. (1908) Hypatia,, in Little Journeys to the Homes of Great Teachers, v. 23 #4 (East Aurora, NY: The Roycrofters), p. 82ff

28. Whitefish, M. T. (2004) *The Ecclesiastical History of Socrates Scholasticus* (Whitefish, MT: Kessinger Publishing Reprint Series)

## Epigenetics and Alma Mahler

### Acknowledgments

Kammerer, Koostler and Whittaker were the subject of a correspondence between my MBL colleagues Porter Anderson and Gary Borisy; this editorial followed their discussion. It was written in the same library that Morgan, Loeb and Whittaker used to sniff out Kammerer's work.

### References

1. Kammerer, P. (1926) Paul Kammerer's letter to the Moscow Academy of Sciences. *Science* **64**, 493–494 [Free Full Text]

2. Vargas, A. O. (2009) Did Paul Kammerer Discover Epigenetic Inheritance? A Modern Look at the Controversial Midwife Toad Experiments. *J. Exp. Zool. B Mol. Dev. Evol.* **312**, 667–678 [Medline]

3. Mahler-Werfel, A. (1960) *Mein Leben* (Gerald Weissmann translation) (Frankfurt: Fischer Verlag), p. 54

4. Gliboff, S. (2006) The Case of Paul Kammerer: Evolution and Experimentation in the Early 20th Century. *J. Hist. Biol.* **39**, 525–563 [CrossRef]

5. Kammerer, P. (1912) *Über Werbung und Vererbung der musikalischen Talents* (Leipzig: Theodor Thomas)

6. Anonymous (1923) Biologist to Tell How Species Alter. *The New York Times,* Nov. 28, 5

7. The Research of Controversial Scientist Paul Kammerer. redOrbit, http://www.redorbit.com/news/science/1747240/the_research_of_controversial_scientist_paul_kammerer/index.html. Posted Sep. 2, 2009

8. Kammerer, P. (1909) Vererbung erzwungener Fortpflanzung sanpassungen. III. Mitteilung: Die Nachkommen der nicht brutpflegenden. *Alytes obstetricans. Archiv. für Entwicklung smechanik der Organismen* **28**, 447–545 [CrossRef]

9. Koestler, A. (1972) *The Case of the Midwife Toad* (New York: Vintage)

10. Gould, S. J. (1972) Zealous Advocates. Book Reviews: The Case of the Midwife Toad. *Science* **176**, 623–625 [Free Full Text]

11. Wagner, G. P. (2009) Paul Kammerer's Midwife Toads: About the Reliability of Experiments and Our Ability to Make Sense of Them. *J. Exper Zool. B Mol. Dev. Evol.* **312**, 665–666

12. Pennisi, E. (2009) History of Science. The Case of the Midwife Toad: Fraud or Epigenetics? *Science* **325**, 1194–1195 [Abstract/Free Full Text]

13. Woolf, B. (1955) On C. H. Waddington's 50th Birthday, cited in Robertson, A. (1977) Conrad Hal Waddington. *Biogr. Mem. Fellows R. Soc.* **23**, 575–622

14. Schoenherr, C. J., Tilghman, S. M. (2000) Epigenetics in Mammals. In Elgin, S. C. R., Workman, J. L. eds. *Chromatin Structure and Gene Expression* (Oxford: Oxford University Press), pp. 253–269

15. Changeux, J. P., Courrège, P., Danchin, A. A. (1973) A Theory of the Epigenesis of Neuronal Networks by Selective Stabilization of Synapses. *Proc. Natl. Acad. Sci. USA* **70**, 2974–2978 [Abstract/Free Full Text]

16. Gliboff, S. J. (2001) The Pebble and the Planet: Paul Kammerer, Ernst Haeckel, and the Meaning of Darwinism, dissertation (Baltimore: Johns Hopkins University), p. 204

17. Keegan, S. (1992) *The Bride of the Wind: The Life and Times of Alma Mahler-Werfel* (New York: Viking), p. 172

18. Lehrer, T. (Recorded July 1965) Alma. *That Was the Year That Was*. Reprise/Warner Bros. Side 2, Track 4

19. Baur, E. (1914) Bemerkungen zu Kammerers Abhandlung: Vererbung erzwungener Farbveränderung IV. *Arch. Entwick.* **38**, 682–684 [CrossRef]

20. Kammerer, P. (1914) Aufklärung zu vorstehenden Bemerkungen des Herrn Professor Baur. *Arch. Entwick.* **38**, 684 [CrossRef]

21. Gliboff, S. (2005) Protoplasm . . . Is Soft Wax in Our Hands: Paul Kammerer and the Art of Biological Transformation. *Endeavour* **29**, 162–167 [CrossRef] [Medline]

22. Online submission information. *The FASEB Journal,* http://www.fasebj.org/misc/onlinesub.shtml

23. Hwang, W. S., Roh, S. I., Lee, B. C., Kang, S. K., Kwon, D. K., Kim, S., Kim, S. J., Park, S. W., Kwon, H. S., Lee, C. K., et al. (2005) Patient-Specific Embryonic Stem Cells Derived from Human SCNT Blastocysts. *Science* **308**, 1777–1783 [Abstract/Free Full Text]

24. Verlet, J. R., Bragg, A. E., Kammrath, A., Cheshnovsky, O., Neumark, D. M. (2005) Comment on "Characterization of Excess Electrons in Water-Cluster Anions by Quantum Simulations." *Science* **310**, 1769 [Free Full Text]

25. Kennedy, D. (2006) Editorial Retraction. *Science* **311**, 335 [Medline]

26. Editorial Expression of Concern for Multiple Articles. *Proc. Natl. Acad. Sci. USA* 2010;107, 6551 [Free Full Text]

27. Retraction. Interaction of Discoidin Domain Receptor 1 Isoform b (DDR1b) with Collagen Activates p38 Mitogen-Activated Protein Kinase and Promotes Differentiation of Macrophages. *FASEB J.* 2009;23, 3251 [Free Full Text]

28. Rossner, M. (2006) How to Guard Against Image Fraud. *The Scientist* **20**, 24–25

29. Noble, G. K. (1926) Kammerer's *Alytes. Nature* **18**, 209–210

30. Morgan, T. H. (1926) Letter to G. K. Noble, cited in Aronson, L. R. (1975) The Case of the Midwife Toad. *Behav. Genet.* **5**, 115–125 [CrossRef]

31. Kammerer, P. (1923) Experiments on *Ciona* and *Alytes. Nature* **112**, 826–827

32. Whittaker, J. R. (Aug. 1985) Paul Kammerer and the Suspect Siphons. *MBL Science,* http://www.mbl.edu/publications/pub_archive/Ciona/Kammerer/index.html

33. Fox, H. M. (1923) Dr. Kammerer's *Ciona* Experiments. *Nature* **122**, 653–654

34. Wermel, J., Lopaschow, G. W. (1930) Über den Einfluss der Regeneration und Überernährung auf die Siphonenlänge bei *Ciona intestinalis* L. Ein Beitrag zu Kammerer's Experientmenten. *Arch. Entwicklungsmech. Org.* **122**, 22–47 [CrossRef]

35. Whittaker, J. R. (1975) Siphon Regeneration in *Ciona. Nature* **255**, 224–225 [CrossRef] [Medline]

36. Loeb, J. (1916) *The Organism as a Whole, from a Physiochemical Viewpoint* (New York: Putnam), p. 339

37. Kammerer, P. (1924) *The Inheritance of Acquired Characteristics* (New York: Boni & Liveright)

38. Kammerer, P. (1912) Körper Kultur und Rasse. *Das oesterreichische Sanitatswesen* **24**, 441–452

## Inflammation Is Complicated: From Metchnikoff to Meryl Streep

1. Dargis, M. (2009) A September-September Romance. *The New York Times,* Dec. 25, p. C1

2. Metchnikoff, E. (1968; orig. 1905) *Immunity in Infective Disease,* vol. 61 (New York: Johnson Reprint Corp.), p. 245ff.

3. Forscher, B. K., Houck, J. C., Thomas, L. (1971) Inflammation as an Adaptive Response. In *Immunopathology of Inflammation,* Forscher, B. K., Houck, J. C., eds. (Amsterdam: Excerpta Medica), pp. 1–12

4. Serhan, C. N., Brain, S. D., Buckley, C. D., Gilroy, D. W., Haslett, C., O'Neill, L. A., Perretti, M., Rossi, A. G., Wallace, J. L. (2007) Resolution of Inflammation: State of the Art, Definitions and Terms. *FASEB J.* 21, 325–332 [Abstract/FREE Full Text]

5. Virchow, R. (1858) *Die Cellularpathologie in ihrer Begründung auf physiologische und pathologische Gewebelehre* (Berlin: August Hirschwald), p. 430

6. Gallin, J. I., Snyderman, R., Cotran, R. S. (1995) Inflammation: Historical Perspectives. In *Inflammation: Basic Principles and Clinical Correlates,* 3rd ed., Gallin, J. I., Snyderman, R., eds. (Philadelphia: Lippincott Williams & Wilkins), pp. 5–10

7. D'Acquisto, F., Maione, F., Pederzoli-Ribeil, M. (2010) From IL-15 to IL-33: The Never-Ending List of New Players in Inflammation. Is It Time to Forget the Humble Aspirin and Move Ahead? *Biochem. Pharmacol.* **79**, 525–534 [CrossRef] [Medline]

8. Qu, Y., Franchi, L., Nunez, G., Dubyak, G. R. (2007) Nonclassical IL-1 β Secretion Stimulated by P2X7 Receptors Is Dependent on Inflammasome Activation and Correlated with Exosome Release in Murine Macrophages. *J. Immunol.* **179**, 1913–1925 [Abstract/FREE Full Text]

9. Boccardo, E., Lepique, A. P., Villa, L. L. (2010) The Role of Inflammation in HPV Carcinogenesis. *Carcinogenesis* [Epub ahead of print] (2010) *Carcinogenesis* **31** (11), 1905–1912

10. Rosenstein, E. D., Greenwald, R. A., Kushner, L. J., Weissmann, G. (2004) Hypothesis: The Humoral Immune Response to Oral Bacteria Provides a Stimulus for the Development of Rheumatoid Arthritis. *Inflammation* **28**, 311–318 [CrossRef] [Medline]

11. Watson, C., Alp, N. J. (2008) Role of Chlamydia Pneumoniae in Atherosclerosis. *Clin. Sci. (Lond.)* 114, 509–531 [Medline]

12. Horace, *Epistles* I.i, 1.13, http://www.thelatinlibrary.com/horace/epist1.shtml

13. Weissmann, G., Smolen, J. E., Korchak, H. M. (1980) Release of Inflammatory Mediators from Stimulated Neutrophils. *N. Engl. J. Med.* **303**, 27–34 [Medline] [Web of Science]

14. Persol, G. M., Borts, E. L., McFarland, J. (1940) Inflammation. In *Cyclopedia of Medicine, Surgery and Specialties,* vol. vii, Persol, G. M., Borts, E. L., eds. (Philadelphia: F.A. Davis), p. 806

15. France, A. (1905) *Sur la pierre blanche* (Paris: Calmann-Lévy), p. 112.

16. Karrer, P. (1958) Paul Ehrlich. *J. Chem. Educ.* **35**, 392–398 [CrossRef]

17. Roitt, I. M., Doniach, D. (1957) Auto-Immunization in Thyroid Diseases. *Proc. R. Soc. Med.* **50**, 958–961 [Medline]

18. Schroder, K., Tschopp, J. (2010) The Inflammasomes. *Cell* **140**, 821–832 [CrossRef] [Medline]

**An *Arrowsmith* for the NASDAQ Era: *Extraordinary Measures***

1. Lewis, S. (1925) *Arrowsmith* (New York: Harcourt, Brace), p. 158

2. Armstrong, D. (2006) Genzyme Drug for Rare Pompe Disease. *The Wall Street Journal,* Apr. 29, p. A2

3. Anand, G. (2006) *The Cure: How a Father Raised $100 Million—and Bucked the Medical Establishment—in a Quest to Save His Children* (New York: HarperCollins), p. 303

4. Stevens, D. (2010) If I Don't Write a Good Business Plan, My Child Will Die. *Slate,* posted Jan. 21, http://www.slate.com/id/2242213/

5. Hornaday, A. (2010) Diagnosis: Predictably Earnest. *The Washington Post* Jan. 22, p. 7

6. Hirschhorn, R., Weissmann, G. (1976) Genetic Disorders of Lysosomes. *Prog. Med. Genet.* **1**, 49–101 [Medline]

7. Hirschhorn, R., Reuser, A. J. (2001) Glycogen Storage Disease Type II: Acid α-glucosidase (acid maltase) Deficiency. Scriver, C. R., Beaudet, A., Sly, W. S., Valle, D. eds. *The Metabolic and Molecular Bases of Inherited Disease* (New York: McGraw-Hill), pp. 3389–3420

8. Sando, G. N., Neufeld, E. F. (1977) Recognition and Receptor-Mediated Uptake of a Lysosomal Enzyme, α-l-iduronidase, by Cultured Human Fibroblasts. *Cell* **12**, 619–627 [CrossRef] [Medline]

9. Van Hove, J. L. K., Yang, H. W., Wu, J-T. Y., Brady, R. O., Chen, Y. T. (1996) High-level Production of Recombinant Human Lysosomal Acid α-Glucosidase in Chinese Hamster Ovary Cells Which Targets to Heart Muscle and Corrects Glycogen Accumulation in Fibroblasts from Patients with Pompe Disease. *Proc. Natl. Acad. Sci. USA* **93**, 65–70 [Abstract/ Free Full Text]

10. Nobelprize.org. Christian de Duve: The Nobel Prize in Physiology or Medicine 1974, Nobel Lecture Dec. 12, 1974, Exploring Cells with a Centrifuge, http://nobelprize.org/ nobelprizes/medicine/laureates/1974/duve-lecture.html

11. Ailworth, E. (2010) Hollywood Treatment: A True Story That Unfolded at Genzyme Inspired Film "Extraordinary Measures." *Boston Globe,* Jan. 16, p. B5

12. O'Donnell, K. (2010) Pompe Disease—The Real Story, blog, last post Feb. 6, http:// pompestory.blogspot.com

13. Shottenkirk, J. (2006) Patience Pays off for William Canfield: Moves up Through Biotech Industry. *Journal Record* (Oklahoma City, OK), Nov. 1, 1

14. Anand, p. 256

15. Anonymous (2006) Pompe Disease, U.S. FDA Approves Genzyme's Myozyme for Patients with Pompe Disease. *Biotech Business Week,* June 1, 304

16. Pompe, J. C. (1936) Cardiomegalia Glycogenica (glycogenic cardiomegaly), Doctoral Thesis. Amsterdam, Dekker & Van de vegt NV. Nijmegen-Utrecht

17. Høyer, S., Samuelson, G. (1986) Mannen bakom syndromet: Joannes Cassianus Pompe. Först att påvisa glykogeninlagring vid hjärtförstoring—fick en våldsam död framför en exekutionspluton. *Läkartidningen, Stockholm* **83**, 1477–1479

18. Die Glycogenose, German website. Selbsthilfegruppe Glykogenase Deutschland e.V., Johann Pompe (1901–1945), http://www.glykogenose.de/index.php?option=comcontent&view=article&id=135:joannes-cassianus-pompe-1901–1945&catid=37:typ-2&Itemid=77

19. James, H. (1999; orig. 1904) *The Golden Bowl* Oxford World's Classics (Oxford: Oxford University Press), p. 13

20. Twort, F. W. (1915) An Investigation on the Nature of Ultra-Microscopic Viruses. *Lancet* **186**, 1241–1243 [CrossRef]

21. d'Hérelle, F. (1917) Sur un microbe invisible antagoniste des bacilles dysentériques. *Acad. Sci. Paris* **165**, 373–375

22. Lewis, p. 122

23. de Kruif, P. (1962) *The Sweeping Wind* (New York: Harcourt Brace), p. 43

24. Anderson, E. T. (1978) Plague in the Continental United States, 1900–76. *Public Health Rep.* **93**, 297–301 [Medline]

25. Williams, C. L. (1935) History of Bubonic Plague in New Orleans. *Am. J. Trop. Med.* **s1–s15**, 555–569

26. Lewis, pp. 328–330

27. Hall, M. (1931) The Screen: A Nobel Prize Novel. *New York Times*, Dec. 8, p. 36

28. Stone, R. (2002) Bacteriophage Therapy: Stalin's Forgotten Cure. *Science* **298**, 728–731 [Free Full Text]

29. Lederberg, J. (1996) Smaller Fleas . . . ad Infinitum: Therapeutic Bacteriophage Redux. *Proc. Natl. Acad. Sci. USA* **93**, 3167–3168 [Abstract/Free Full Text]

30. Phage International. Commercial Companies Involved in Bacterio Phage Research and Development, http://www.phageinternational.com/phagetherapy/companies.htm

31. Uhr, J. W., Weissmann, G. (1965) Intracellular Distribution and Degradation of Bacteriophage in Mammalian Tissues. *J. Immunol.* **94**, 544–550 [Abstract/Free Full Text]

32. Merril, C. R., Scholl, D., Adhya, S. L. (2003) The Prospect for Bacteriophage Therapy in Western Medicine. *Nat. Rev. Drug Discov.* **2**, 489–497 [CrossRef] [Medline]

## Sarah Palin and Marie-Antoinette: Post-Traumatic Tress Disorder

1. Navarini, A. A., Nobbe, S., Trüeb, R. M. (2009) Marie Antoinette Syndrome. *Arch. Dermatol.* **145**, 656 [Free Full Text]

2. Cooper, H. (2009) 44 Days in the White House, and the Hair? Grayer Already. *The New York Times,* Mar. 5 http://query.nytimes.com/gst/fullpage.html?res= 9801EFD6133DF936A3

5750C0A96F9C8B63&scp=1&sq=44%20Days%20in%20the%20White%20House&st=cse. Accessed Aug. 18, 2009

3. Rutenburg, J., Kovalesky, S. F. (2009) Retracing Palin's Long March to a Short-Notice Resignation. *The New York Times,* July 13 http://www.nytimes.com/2009/07/13/us/politics/13palin.html?r=1&scp=1&sq=Retracing%20Palin's%20Long%20March%20to%20a%20Short-Notice%20Resignation&st=cse. Accessed Aug. 18, 2009

4. Hosford, D. (2004) The Queen's Hair: Marie-Antoinette, Politics, and DNA. *Eighteenth-Century Studies* **38,** 183–200.

5. Teitell, B. (2009) The Palin Debate: Tress Stress? Experts at Odds About Cause of Thinning Hair. *Boston Globe*, July 23, G22

6. Teitell, B. (2009) Sarah Palin's Hairdresser Speaks. *Boston Globe*, July 23, G22

7. Dowd, M. (2009) Sarah Grabs the Grievance Grab Bag from Hillary. *The New York Times,* July 29, http://www.nytimes.com/2009/07/29/opinion/29dowd.html?scp=1&sq=Sarah%20grabs%20the%20grievance%20grab%20bag%20from%20Hillary&st=cse. Accessed Aug. 18, 2009

8. Jelinek, J. E. (1972) Sudden Whitening of the Hair. *Bull. N.Y. Acad. Med.* **48**, 1003–1013 [Medline]

9. Skellett, A.-M., Millington, G. W. M., Levell, N. J. (2008) Sudden Whitening of the Hair: An Historical Fiction? *J. R. Soc. Med.* **101**, 574–576 [Free Full Text]

10. Fraser, A. (2002) *Marie-Antoinette: The Journey* (New York: Anchor Books), p. 162

11. Ibid., p. 344

12. Goncourt, E., Goncourt, K. J. (1858) *Histoire de Marie-Antoinette* (Paris: Firmin-Didot) p. 419 (quoted in ref. 8)

13. Carlyle, T. (2002; orig. 1837) Rosenberg, J., ed. *The French Revolution: A History* (New York: Modern Library), p. 419

14. Fraser, p. 404

15. Montagna, W., cited by Harrison, S., Sinclair, R. (2002) Telogen Effluvium. *Clin. Exp. Dermatol.* **27**, 389–395 [CrossRef] [Medline]

16. Ohyama, M. J. (2007) Hair Follicle Bulge: A Fascinating Reservoir of Epithelial Stem Cells. *Dermatol. Sci.* **46**, 81–89 [CrossRef]

17. Wasserman, D., Guzman-Sanchez, D. A., Scott, K., McMichael, A. (2007) Alopecia Areata. *Int. J. Dermatol.* **46**, 121–131 [Medline]

18. Arck, P. C., Overall, R., Spatz, K., Liezman, C., Handjiski, B., Klapp, B. F., Birch-Machin, M. A., Peters, E. M. J. (2006) Towards a "Free Radical Theory of Graying": Melanocyte Apoptosis in the Aging Human Hair Follicle Is an Indicator of Oxidative Stress Induced Tissue Damage. *FASEB J.* **20**, 1567–1569 [Abstract/Free Full Text]

19. Wood, J. M., Decker, H., Hartmann, H., Chavan, B., Rokos, H., Spencer, J. D., Hasse, S., Thornton, M. J., Shalbaf, M., Paus, R., Schallreuter, K. U. (2009) Senile Hair Graying: H2O2-Mediated Oxidative Stress Affects Human Hair Color by Blunting Methionine Sulfoxide Repair. *FASEB J.* **23**, 2065–2075 [Abstract/Free Full Text]

20. Parker-Pope, T. (2009) Unlocking the Secrets of Gray Hair. *The New York Times,* Mar. 10 http://www.nytimes.com/2009/03/10/health/10well.html?scp=1&sq=Unlocking%20the%20Secrets%20of%20Gray%20Hair&st=cse. Accessed Aug. 18, 2009

21. Arck, P. C., Handjiski, B., Hagen, B., Joachim, R., Klapp, B. F., Paus, R. (2001) (Indications for a Brain-Hair Follicle Axis: Inhibition of Keratinocyte Proliferation and Up-Regulation of Keratinocyte Apoptosis in Telogen Hair Follicles by Stress and Substance P. Sep. 17, *FASEB J.* doi:10.1096/00–0699fje

22. Telek, A., Bíró, T., Bodó, E., Tóth, B. I., Borbíró, I., Kunos, G., Paus, R. (2007) Inhibition of Human Hair Follicle Growth by Endo- and Exoc Annabinoids. *FASEB J.* **21**, 3534–3541 [Abstract/Free Full Text]

23. Peters, E. M., Liotiri, S., Bodó, E., Hagen, E., Bíró, T., Arck, P. C., Paus, R. (2007) Probing the Effects of Stress Mediators on the Human Hair Follicle: Substance P Holds Central Position. *Am. J. Pathol.* **171**, 1872–1886 [Abstract/Free Full Text]

## Coca-Cola and H. G. Wells: Dietary Supplements as Subprime Drugs

1. Ledbetter, D. O. (2008) Penalized Tackle Hits Pill Maker with Suit. *Atlanta Journal-Constitution*, Nov. 12, D1

2. United Press International (2008) Press Release: Starcaps-Brand Dietary Supplement Recalled http://www.upi.com/Science_News/2008/12/10/Starcaps-brand_dietary_supplement_recalled /UPI-66641228933811/. Last accessed Mar. 23, 2009

3. U.S. Food and Drug Administration, Center for Drug Evaluation and Research (2009) Consumer Directed Questions and Answers about FDA's Initiative Against Contaminated Weight Loss Products. http://www.fda.gov/Cder/consumerinfo/weight_loss_products.htm. Last accessed Mar. 23, 2009

4. Wells, H. G. (2003; 1909) *Tono-Bungay* (New York: Modern Library Classics/ Random House; London: Duffield & Co.), p. 93

5. Singer, N. (2009) FDA Finds "Natural" Diet Pills Laced with Drugs. *The New York Times*, Feb. 9, http://www.nytimes.com/2009/02/10/business/10pills.html. Last accessed Mar. 23, 2009

6. Allen, F. (1994) *Secret Formula: How Brilliant Marketing and Relentless Salesmanship Made Coca-Cola the Best-Known Product in the World* (New York: Harper Business), p. 42

7. Wells, *Tono-Bungay,* p. 150

8. Anonymous (1909) Patent Medicine in a Novel Form; H. G. Wells, in "Tono-Bungay," Expands a Prosaic Theme into Highly Entertaining Fiction. *The New York Times*, Jan. 30, BR54

9. Kemp, K. W. (2002) *God's Capitalist: Asa Candler of Coca-Cola* (Macon, GA: Mercer University Press), p. 56

10. Wells, *Tono-Bungay*, p. 256

11. James, W. (1920) *The Letters of William James,* James, H. ed. (New York and Boston: Atlantic Monthly Press), vol. 2, p. 316

12. DietSpotlight.com (ND) Star Caps Review, http://www.dietspotlight.com/star-caps-review/. Last accessed Mar. 23, 2009

13. Meolie, A. L., Rosen, C., Kristo, D., Kohrman, M., Gooneratne, N., Aguillard, R. N., Fayle, R., Troell, R., Townsend, D., Claman, D., Hoban, T., Mahowald, M. Clinical Practice Review Committee, American Academy of Sleep Medicine (2005) Oral Nonprescription Treatment for Insomnia: An Evaluation of Products with Limited Evidence. *J. Clin. Sleep Med.* **1**, 173–187 [Medline]

14. Grabovac, V., Schmitz, T., Föger, F., Bernkop-Schnürch, A. (2007) Papain: An Effective Permeation Enhancer for Orally Administered Low Molecular Weight Heparin. *Pharm. Res.* **24**, 1001–1006 [CrossRef] [Medline]

15. Durlach, J., Durlach, V., Bac, P., Bara, M., Guiet-Bara, A. (1994) Magnesium and Therapeutics. *Magnes. Res.* **7**, 313–328 [Medline]

16. U.S. Food and Drug Administration (1995) Guide to Nutrition Labeling and Education Act (NLEA) Requirements (August 1994) http://www.fda.gov/ora/inspect_ref/igs/nleatxt.html. Last accessed Mar. 23, 2009

17. Saper, R. B., Phillips, R. S., Sehgal, A., Khouri, N., Davis, R. B., Paquin, J., Thuppil, V., Kales, S. N. (2008) Lead, Mercury, and Arsenic in US- And Indian-Manufactured Ayurvedic Medicines Sold Via the Internet. *JAMA* **300**, 915–923 [Abstract/Free Full Text]

18. Poortmans, J. R., Francaux, M. (1998) Renal Dysfunction Accompanying Oral Creatine Supplements. *Lancet* **352**, 234 [Medline]

19. American Botanical Council (1994) Congress Passes Dietary Supplement Health and Education Act of 1994: Herbs to Be Protected as Supplements. *HerbalGram* **32**, 18 http://content.herbalgram.org/iherb/herbalgram/articleview.asp?a=1114&p=Y. Last accessed Mar. 23, 2009

20. Herrera, S., Bruguera, M. (2008) Hepatotoxicity Induced by Herbs and Medicines Used to Induce Weight Loss. *Gastroenterol Hepatol.* **31**, 447–453 [CrossRef] [Medline]

21. Violon, C. J. (1997) Belgian (Chinese Herb) Nephropathy: Why? *J. Pharm. Belg.* **52**, 7–27 [Medline]

22. U.S. Food and Drug Administration *Dietary Supplement Health and Education Act of 1994, Public Law 103–417, 103rd Congress* http://www.fda.gov/opacom/laws/DSHEA.html. Last accessed Mar. 23, 2009

23. Wells, *Tono-Bungay,* p. 159

24. Wallace, R. B., Gryzlak, B. M., Zimmerman, M. B., Nisly, N. L. (2008) Application of FDA Adverse Event Report Data to the Surveillance of Dietary Botanical Supplements. *Ann. Pharmacother.* **42**, 653–660 [Abstract/Free Full Text]

25. Ries, C. A., Sahud, M. A. (1975) Agranulocytosis Caused by Chinese Herbal Medicines. Dangers of Medications Containing Aminopyrine and Phenylbutazone. *JAMA* **231**, 352–355 [Abstract/ Free Full Text]

26. Wells, *Tono-Bungay,* pp. 251–257

**Voodoo Economics and Voodoo Healing:**
**Witchcraft Persists in Massachusetts**

1. Bearak, B. (2009) Pope Tells Clergy in Angola to Work Against Belief in Witchcraft. *The New York Times*, Mar. 22, p. A8

2. Penczak, C. (2005) Using Reiki Magick. *Llewellyn Journal,* Jan. 10, http://www.llewellynjournal.com/article/751. Accessed Apr. 2009

3. Dana-Farber Cancer Institute, Cancer Information, Integrative Therapies: Reiki. http://dana-farber.org/can/integrative-therapies/html/reiki.html. Accessed Apr. 2009

4. Bush, G. H. W. (1980) Speech at Carnegie Mellon University, Apr. 10, NBC Nightly News (online files), http://icue.nbcunifiles.com/icue/files/icue/site/pdf/33292.pdf. Accessed Apr. 2009

5. Pullela, P. (2009) Shun Witchcraft, Pope Tells Angolan Catholics. Reuters India, Mar. 21, http://in.reuters.com/article/worldNews/idINIndia-38634620090321. Accessed Apr. 2009

6. Stickler, A. (2005) Angola Witchcraft's Child Victims. BBC News, July 13, http://news.bbc.co.uk/2/hi/africa/4677969.stm. Accessed Apr. 2009

7. Salopek, P. (2004) Children in Angola Tortured As Witches by Own Families. Chicago Tribune.com, Mar. 28, http://www.chicagotribune.com/services/newspaper/printedition/sunday/chi-0403280349mar28,0,2660526.story. Accessed Apr. 2009

8. U.S. Central Intelligence Agency (2009) https://www.cia.Gov/library/publications/the-world-factbook/geos/ao.html. Accessed Apr. 2009

9. World Bank (2006) Annual Report 2006: Country Eligibility for Borrowing from the World Bank Annual Report 2006: Country Eligibility for Borrowing from the World Bank http://web.worldbank.org/WBSITE/EXTERNAL/EXTABOUTUS/EXTANNREP/EXTAN NREP2K6/0,,contentMDK:21046862~menuPK:2915976~pagePK:64168445~piPK:64168 309~theSitePK:2838572,00.html. Accessed Apr. 2009

10. Witches of the World (2009) Massachusetts Home Page, http://www.witchvox.com/vn/hm/usma.html. Accessed Apr. 2009

11. Harvard Medical School (October 2005) Healing Touch Therapy: Alternative Therapies Relax Heart Patients. *Harvard Health Letter* https://www.health.harvard.edu/press_releases/healing_touch_therapy. Accessed Apr. 2009

12. Plymouth Are Witches Healing and Divination Group (2009) Welcome, Plymouth Area Witches! http://www.meetup.com/Plymouth-Area-Witches/. Accessed Apr. 2009

13. Masson, D. The Salgion Tradition: A Brief History, http://www.circleofsalgion.org/history.html. Accessed Apr. 2009

14. Kerr, C. E., Wasserman, R. H., Moore, C. I. (2007) Cortical Dynamics as a Therapeutic Mechanism for Touch Healing. *J. Altern. Complementary Med.* **13**, 59–66 [CrossRef] [Medline]

15. National Institutes of Health, NCCAM (2008) Reiki: An Introduction, http://nccam.nih.gov/health/reiki/. Accessed Apr. 2009

16. Brown, J. (2005) Witches Work Their Magic in Hurricane Zone: Pagan Rituals Help Some Recover from Katrina. ABC News, Oct. 31, http://abcnews.go.com/GMA/story?id=1256677. Accessed Apr. 2009

17. The Witches' Voice (2009) Group Profile: Provincetown MA aka P-Town Witch School, http://www.witchvox.com/vn/vn_detail/dt_gr.html?a=usma&id=33687. Accessed Apr. 2009

18. The Bead Tree with JoAnn Allard, http://www.thebeadtree.com/Magical_ Mixtures.htm. Accessed Apr. 2009

19. Cape Cod Body and Soul Day Spa, http://www.capecodbodyandsoul.com/who.htm. Accessed Apr. 2009

20. Lowell, J. R. (1870) Witchcraft. *Literary Essays* (Boston:Houghton Mifflin), p. 331ff

21. Rosenthal, B. (1993) *Salem Story: Reading the Witch Trials of 1692* (Cambridge and New York: Cambridge University Press), p. 111ff

22. Weissmann, G. (1993) Commemoration and Witchcraft. *Democracy and DNA* (New York: Hill & Wang), p. 125

23. Fogel, R. W. (2004) *The Escape from Hunger and Premature Death: Europe, America, and the Third World* (Cambridge: Cambridge University Press), p. 36

24. National Institutes of Health, NCCAM (2003) Questions and Answers About Homeopathy, http://nccam.nih.gov/health/ homeopathy/. Accessed Apr. 2009

25. Assefi, N., Bogart, A., Goldberg, J., Buchwald, D. (2008) Reiki for the Treatment of Fibromyalgia: A Randomized Controlled Trial. *J. Altern. Complement Med.* **14**, 1115–1122 [CrossRef] [Medline]

## Myrna Loy: Co-Principal Investigator

1. Hammett, D., Hackett, A., Goodrich, F. (1939) *Another Thin Man* (MGM film)

2. Handelsman, J., Grymes, R. A. (2008) Looking for a Few Good Women? *DNA and Cell Biol.* **27**, 463–465 [CrossRef] [Medline]

3. French, P. (2008) Screen legends: No 17. Myrna Loy 1905–93. *Observer* (London), May 18, 14

4. Hellman, L. (1976). *Scoundrel Time* (Boston: Little, Brown), p. 69

5. Kristol, W. (2008) Both Sides Now. *The New York Times,* Sep. 14, A25

6. AP (2008) Ferraro Quits Clinton Post After Remarks About Obama, Mar. 13

7. The Rosalind Franklin Society (home page), http://www.rosalindfranklinsociety.org. Accessed Sep. 2008

8. The Ellison Medical Foundation (home page), http://www.ellisonfoundation.org. Accessed Sep. 2008

9. Martinson, D. (2005) *Lillian Hellman: A Life with Foxes and Scoundrels* (New York: Counterpoint), p. 92

10. Mellen, J. (1996) *Hellman and Hammett: The Legendary Passion of Lillian Hellman and Dashiell Hammett* (New York: Harper Collins)

11. O'Hara, F. (1957) "To the Film Industry in Crisis," *Meditations in an Emergency* (New York: Grove Press), p. 34

12. Maslin, J. (1985) Tribute to Myrna Loy. *The New York Times*, Jan. 16, C17

13. Folkart, B. A. (1993) Myrna Loy, Star of "Thin Man" Films, Dies at 88. *Los Angeles Times*, Dec. 15, 3

14. Hammett, D., et al. *The Thin Man*

15. Hammett, D. (1934) *The Thin Man* 1992 ed. (New York: Vintage Books), p. 150

16. Anon (1961) Dashiell Hammett Dies, Created Hard-Boiled Detectives . . . Jailed for Aiding Reds. *The New York Times*, Jan. 11, 47

17. Leonard, J. (2005) New Books. *Harper's Magazine* **311**, 85–87

18. Anon. Samuel Dashiel Hammett: Sergeant, United States Army—Author. Arlington National Cemetery Website http://www.arlingtoncemetery.net/shammett.htm. Accessed Sep. 2008

## Dr. Ehrlich and Dr. Atomic: Beauty vs. Horror in Science

1. Overbye, D. (2008) "Doctor Atomic" at the Met: The Terror and Attraction of Science, Put to Song. *The New York Times,* Oct. 18, D2

2. Falkow, S. (2008) I Never Met a Microbe I Didn't Like. *Nat. Med.* **14**, 1053–1057 [CrossRef] [Medline]

3. Hitchens, C. (2008) Sarah Palin's War on Science: The GOP Ticket's Appalling Contempt for Knowledge and Learning. *Slate,* Oct. 27, http://www.slate.com/id/2203120/. Accessed Nov. 2008

4. Cohen, S. (2008) Award Presentation, Lasker-Koshland Special Achievement Award in Medical Science. The Lasker Foundation. http://www.laskerfoundation.org/awards/2008_s_presentation.htm. Accessed Nov. 2008

5. Berg, P., Baltimore, D., Boyer, H. W., Cohen, S. N., Davis, R. W., Hogness, D. S., Nathans, D., Roblin, R., Watson, J. D., Weissman, S., Zinder, N. D. (1974) Letter: Potential Biohazards of Recombinant DNA Molecules. *Science* **185**, 303 [Free Full Text]

6. Marquardt, M. (1951) *Paul Ehrlich* (New York: Schuman)

7. The Nobel Foundation. http://nobelprize.org/nobel_prizes/medicine/laureates/1908/index.html. Accessed Nov. 2008

8. The Lasker Foundation Albert Lasker Medical Research Awards in Basic Science. http://www.laskerfoundation.org/awards/2008 basic.htm. Accessed Nov. 2008

9. Leyden, J. G. (1999) From Nobel Prize to Courthouse Battle; Paul Ehrlich's "Wonder Drug" for Syphilis Won Him Acclaim but Also Led Critics to Hound Him. *The Washington Post,* July 27, Z16

10. Deutsches Historisches Museum, Berlin. Biographie: Paul Erlich, 1854–1915. http://www.dhm.de/lemo/html/biografien/EhrlichPaul. Accessed Nov. 2008

11. Ross, A. (2008) "Doctor Atomic" at the Met. *The New Yorker,* Oct. 27, p. 92

12. Donne, J. (~1609) Holy Sonnet XIV. *Poems of John Donne,* Chambers, E. K., ed., 1896 (London: Lawrence & Bullen), vol. I, p. 16

13. Jungk, R. (1958) *Brighter Than a Thousand Suns: A Personal History of the Atomic Scientists* (Cleugh, J. trans.) (New York: Harcourt Harvest)

14. Ibid., p. 296

15. Oppenheimer: A Life (2004). J. Robert Oppenheimer Centennial at Berkeley, University of California, Berkeley. http://ohst.berkeley.edu/oppenheimer/exhibit/. Accessed Nov. 2008

16. Kahn, H. (1960) *On Thermonuclear War* (Princeton: Princeton University Press)

17. Samuels, G. (1953) A Plea for "Candor" About the Atom: An Eminent Authority on Nuclear Weapons, Dr. J. Robert Oppenheimer Assesses the Wisdom of Letting People Have the Facts. *The New York Times,* June 21, SM8

18. Anonymous (1954) 2 Queries Vital to Loyalty Issue, *The New York Times,* Apr. 13, 20

19. Stern, P. M. (1969) *The Oppenheimer Case: Security on Trial* (New York: Harper & Row)

20. Anonymous (1954) List of Persons Mentioned in Case. Special to *The New York Times. The New York Times,* Apr. 13, 17

21. In the Matter of J. Robert Oppenheimer. Texts of Principal Documents and Letters of Personnel Security Board, General Manager, Commissioners. Washington, DC, May 27, 1954 Through June 29, 1954. U.S. Government Printing Office, Washington, DC. As cited in: The Beat Begins: America in the 1950s. University of Maryland, College Park, Maryland. http://honors.umd.edu/HONR269J/archive/AEC540629.html. Accessed Nov. 2008

22. Wald, G. (1969) A Generation in Search of a Future (speech delivered at the Massachusetts Institute of Technology), Mar. 4. http://www.elijahwald.com/generation.html. Accessed Nov. 2008

23. Brecht, B. (1943) Willett, J., Manheim, R. eds. *Life of Galileo,* 1994 ed., 108–109 (New York: Arcade), Scene 14

**Free Radicals Can Kill You: Lavoisier and the Oxygen Revolution**

1. Djerassi, C., Hoffmann, R. (2001) *Oxygen* (Weinheim, Germany: Wiley-VCH Verlag), p. 119

2. Lavoisier, A-L. (1775, published 1778) *Mémoire sur la nature du Principe qui se combine avec les Métaux pendant leur calcination et qui en augmente le poids. Mémoires de l'Académie Royale des Sciences Paques,* pp. 520–526

3. Priestley, J. (1775) *Experiments and Observations on Different Kinds of Air* (Birmingham, AL: Thomas Pearson), vol. 2, p. 241

4. Harman, D. (1956) Aging: A Theory Based on Free Radical and Radiation Chemistry. *J. Gerontol.* **11**, 298–300 [Free Full Text]

5. Gerschman, R., Gilbert, D. L., Nye, S. W., Dwyer, P., Fenn, W. O. (1954) Oxygen Poisoning and X-Irradiation: A Mechanism in Common. *Science* **119**, 623–626

6. Harman, D. (2009) Origin and Evolution of the Free Radical Theory of Aging: A Brief Personal History, 1954–2009. *Biogerontology* [Epub ahead of print] (2009) *Biogerontology* **10** *(6),* 773–781

7. Weissman, G. (2008) Claude Bernard and Judah Folkman: Nothing Makes Sense in Medicine Except in the Light of Biology. *FASEB J.* **22**, 943–946 [Free Full Text]

8. Thomas, Reuters *Web of Science* http://thomsonreuters.com/products_services/science/science_products/a-z/web_of_science. Last accessed Jan. 25, 2010

9. *The FASEB Journal* (article search), http://www.fasebj.org/search.dtl. Last accessed Jan. 25, 2010

10. McCord, J. M., Fridovich, I. (1969) The Utility of Superoxide Dismutase in Studying Free Radical Reactions. I. Radicals Generated by the Interaction of Sulfite, Dimethyl Sulfoxide, and Oxygen. *J. Biol Chem.* **244**, 6056–6063 [Abstract/Free Full Text]

11. Goldstein, I. M., Kaplan, H. B., Edelson, H. S., Weissmann, G. (1979) Ceruloplasmin. A Scavenger of Superoxide Anion Radicals. *J. Biol. Chem.* **254**, 4040–4045 [Free Full Text]

12. Briggs, R. T., Drath, D. B., Karnovsky, M. L., Karnovsky, M. J. (1975) Localization of NADH Oxidase on the Surface of Human Polymorphonuclear Leukocytes by a New Cytochemical Method. *J. Cell Biol.* **67**, 566–586 [Abstract/Free Full Text]

13. Lavoisier, A-L. (1952) *Traité Élémentaire de Chimie* (1789), illustrated by Mme. Lavoisier. Hutchins, R. M. Adler, M. J., eds., *Lavoisier, Fourier, Faraday: Great Books of the Western World series,* 1–160, *Encyclopædia Britannica,* see Lavoisier, A. L.

14. Eagle, C. T., Sloan, J. (1998) Marie Anne Paulze Lavoisier: The Mother of Modern Chemistry. *Chem. Educ.* **3**, 1–18

15. Hoffmann, R. (2002) Mme Lavoisier. *Am. Sci.* **90**, 22–24

16. Pinault-Sörensen, M. (1994) Madame Lavoisier, Illustrator and Painter. *La Revue #06* Musée des Arts et Métiers Paris, France. http://www.arts-et-metiers.net/musee.php?P=157&id =10299&lang=ang&flash=f. Last accessed Jan. 25, 2010

17. Guerlac, H. (1975) *Antoine-Laurent Lavoisier, Chemist and Revolutionary* (New York: Charles Scribner's)

18. van Klooster, H. S. (1946) Franklin and Lavoisier. *J. Chem. Educ.* **23**, 107–109 [CrossRef]

19. Scheele, C. W. (1777) *Chemische Abhandlung von der Luft und dem Feuer. Nebst einem Vorbericht von Torbern Bergman* (Uppsala and Leipzig: Magn. Swederus . . . zu finden bey S. L. Crusius)

20. Severinghaus, J. W. (2003) Fire-Air and Dephlogistication. Revisionisms of Oxygen's Discovery. *Adv. Exp. Med. Biol.* **543**, 7–19 [Medline]

21. Cohen, I. B. (1985) *Revolution in Science* (Cambridge, MA: Belknap, Harvard University Press), p. 231

22. McKie, D. (1952) *Antoine Lavoisier: Scientist, Economist, Social Reformer* (New York: Schumann), p. 242

23. Grimaux, E. (1896) *Lavoisier, 1743–1794, d'après sa correspondance, ses manuscrits, ses papiers* (Paris: Cuchet), p. 376

24. Hartley, H. (1947) Antoine Laurent Lavoisier. *Proc. Royal Soc. London, Series B* **134**, 348–377 [Free Full Text]

## Experimental Errors: Paul Bert and the Alabama Tenure Killings

1. Wheaton, S., Dewan, S., Brown, R. (2010) Professor Said to Be Charged After 3 Are Killed in Alabama, *The New York Times*, Feb. 13, A12

2. Dejours, P., Dejours, S. (1992) The Effects of Barometric Pressure According to Paul Bert: The Question Today. *Int. J. Sports Med.* **13** (Suppl. 1), S1–S5 [CrossRef]

3. Special Correspondent (1875) Fatal Ballooning.; The Sad Story of the Zenith. Two Aeronauts Suffocated Above the Clouds—The Survivor's Experience—What Has Been Gained by the Journey—Past Ascensions—Popular Sympathy for the Martyrs to Science. The Fatal Expedition. M. Tissandier's Story. The Return to Earth. Teachings of the Voyage. Aid for the Bereft—The Dead, *The New York Times*, May 2

4. UAHuntsville, Publications, Abstracts, & Presentations. http://www.uah.edu/biology/ amy/publications.html

5. Bishop, A., Gallop, P. M., Karnovsky, M. L. (1998) Pyrroloquinoline Quinone: A Novel Vitamin? *Nutrition Reviews* **56**, 287–294 [Medline]

6. Huntsville Times staff (2010) Why Did Amy Bishop Snap? A Picture of a Driven Woman with a Troubled Past Emerges, *Huntsville Times*, Feb. 22

7. Zezima, K., Dewan, S. (2010) Professor Has No Memory of Shootings, Lawyer Says, *The New York Times*, Feb. 20, A16

8. Taranto, J. (2010) Going Postdoctoral: The Bizarre Case of Prof. Amy Bishop. *The Wall Street Journal online*. http://online.wsj.com/article/SB10001424052748703315004575075073361514318850.html. Accessed Feb. 18, 2010

9. Herring, C., Levitz, J. (2010) Alabama Suspect Had Erratic History. *The Wall Street Journal*, Feb. 17, A3

10. Longfellow, H. W. (1993; orig. 1838) The psalm of life. Hollander, J. ed. *American Poetry: The Nineteenth Cantury* (New York: Library of America), vol. 1, p. 370

11. Scribd Dr. Amy Bishop's Curriculum vitae. http://www.scribd.com/doc/27012456/Dr-Amy-Bishop-s-Curriculum-Vitae

12. Karnovsky, M. L., Bishop, A., Camerero, V. C., Paz, M. A., Colepicolo, P., Ribeiro, J. M., Gallop, P. M. (1994) Aspects of the Release of Superoxide by Leukocytes, And a Means by Which This Is Switched Off. *Environ. Health Perspect.* **102** (Suppl. 10), 43–44

13. Killgore, J., Smidt, C., Duich, L., Romero-Chapman, N., Tinker, D., Reiser, K., Melko, M., Hyde, D., Rucker, R. B. (1989) Nutritional Importance of Pyrroloquinoline Quinone. *Science* **245**, 850–852 [Abstract/Free Full Text]

14. Bishop, A., Marquis, J. C., Cashman, N. R., Demple, B. (1999) Adaptive Resistance to Nitric Oxide in Motor Neurons. *Free Radic. Biol. Med.* **26**, 978–986 [CrossRef] [Medline]

15. Anderson, L. B., Anderson, P. B., Anderson, T. B., Bishop, A. et al. (2009) Effects of Selective Serotonin Reuptake Inhibitors on Motor Neuron Survival. *Int. J. Gen. Med.* May 2009,2 Dovepress.com, http://www.dovepress.com/effects-of-selective-serotonin-reuptake-inhibitors-on-motor-neuron-sur-peer-reviewed-article-IJGM. Accessed Mar. 3, 2010

16. Sweet, L. J., Kantor, I. (2010) Amy Bishop, Husband Listed Kids on Research Paper, *Boston Herald*, Feb. 19, 5

17. Glaisher, J. (1871) *Travels in the Air* (Philadelphia: J. B. Lippincott)

18. Eckener, H. (1938) *Count Zeppelin: The Man and His Work*, Fanell, L. trans. (London: Massie London) p. 50ff

19. Rostène, W. (2006) Paul Bert: Homme de science, homme politique. *J. Société de Biologie* **200**, 245–250 [CrossRef]

20. Tissandier, G. (1875) Le voyage à grande hauteur du ballon "le Zenith." *La Nature* (Paris 3) 337–344

21. Bert, P. (1878) Hitchcock, M. A. Hitchcock, F. A. eds. *La pression barométrique: Recherches de physiologie expérimentale,* translated as *Researches in Experimental Physiology* (1943) (Columbus, OH: College Book Co.), p. 963ff

22. Keith, C. P. (July–Sep. 2005) Catholicisme, bouddhisme et lois laïques au Tonkin (1899–1914) Vingtième Siècle. *Revue d'histoire,* No. 87, *Numéro spécial: Laïcité, séparation, sécularisation 1905–2005*, pp. 113–128

23. Parker-Pope, T. (2010) Genius, Madness and Tenure, *The New York Times* (Well blog) http://well.blogs.nytimes.com/2010/02/22/genius-madness-and-tenure/?apage=5#comments. Accessed Feb. 20, 2010

24. Purcell, A. (2010) When the Shooters Came to Starbucks, *Herald* (Scotland) Mar. 12, 1 http://www.heraldscotland.com/news/world-news/when-the-shooters-came-to-starbucks-1.1013198

25. McKinley, J. C., Jr. (2010) Texas Conservatives Win Curriculum Change, *The New York Times*, Mar. 12, A10

## Monumental Revolutions: Scientific, Sanitary and 'Omic

1. Butterfield, H. (1949) *The Origins of Modern Science* (London: Bell), p. viii

2. Montaigne (ca. 1580) "Of Cannibals," *The Complete Works of Montaigne (1947)* Frame, D., trans, (Stanford, CA: Stanford University Press), p. 150

3. Anon (1859) *Proceedings and Debates of the Third National Quarantine and Sanitary Convention, Issue 9* (New York: Edmund Jones), p. 241

4. American Museum of Natural History (2001) *The Genomic Revolution* (exhibit description). http://www.amnh.org/exhibitions/genomics/0_home/index.html. Accessed Sep. 2009

5. Ubiquitin Drug Discovery and Diagnostics (2009) (program brochure) http://www.ubiquitinconference.com. Accessed Sep. 2009

6. *OMICS: A Journal of Integrative Biology* (overview) http://www.liebertpub.com/products/product.aspx?pid=43. Accessed Sep. 2009

7. Lederberg, J., McCray, A. T. (2001) 'O'me Sweet 'Omics: A Genealogical Treasury of Words. *The Scientist* **15**, 8

8. Yeang, C. H., McCormick, F., Levine, A. J. (2008) Combinatorial Patterns of Somatic Gene Mutations in Cancer. *FASEB J* **22**, 2605–2622 [Abstract/Free Full Text]

9. Davidson, E. H., Levine, M. S. (2008) Properties of Developmental Gene Regulatory Networks. *Proc. Natl. Acad Sci. USA* **105**, 20063–20066 [Abstract/Free Full Text]

10. Trifonov, V., Khiabanian, H., Greenbaum, B., Rabadan, R. (2009) The Origin of the Recent Swine Influenza A(H1N1) Virus Infecting Humans. *Euro. Surveill.* **14**, pii: 19193

11. Hood, L., Ideker, T., Galitski, T. (2001) A New Approach to Decoding Life: Systems Biology. *Ann. Rev. Genom. Human Genetics.* **2**, 343–372 [CrossRef]

12. Kohn, K. W., Aladjem, M. I., Kim, S., Weinstein, J. N., Pommier, Y. (2006) Depicting Combinatorial Complexity with the Molecular Interaction Map Notation. *Mol. Syst. Biol.* **2**, 51 [Medline]

13. Evans, G. A. (2000) Designer Science and the "Omic" Revolution. *Nat. Biotechnol.* **18**, 127 [CrossRef] [Medline]

14. Butterfield, H. (1957) *The Origins of Modern Science* (New York: Free Press)

15. Cohen, I. B. (1985) *Revolution in Science* (Cambridge, MA: Belknap/Harvard University Press)

16. Kuhn, T. S. (1962) *The Structure of Scientific Revolutions* (Chicago: University of Chicago Press)

17. Fleming, D. (1950) *John William Draper and the Religion of Science* (Philadelphia: University of Pennsylvania Press), p. 129

18. Bruno, G. (1585) *Gli Eroici* Project Gutenberg http://ia311009.us.archive.org/0/items/theheroicenthusi19833gut/19833-h/19833-h.htm. Accessed Sep. 2009

19. Cohen, *Revolution in Science*, p. 15

20. Christie, R. C. (1880) *Étienne Dolet: The Martyr of the Renaissance: A Biography* (London: Macmillan), p. 450ff

21. Dolet, E. (1540) *La manière de bien traduire d'une langue en aultre.* Project Gutenberg http://www.gutenberg.org/etext/19483. Accessed Sep. 2009

22. Cohen, M. (1993) *Profane Illumination: Walter Benjamin and the Paris of Surrealist Revolution* (Berkeley, CA: University of California Press), p. 87ff

23. Weissmann, G. (1990) Haussmann on Missiles. *The Doctor with Two Heads,* (New York: Alfred Knopf), p. 128

24. Anonymous (2009) *Piazza di Campo dei Fiori: A Rome Art Lover's Web page.* http://www.romeartlover.it/Vasi28.html. Accessed Sep. 2009

25. Rosenzweig, R., Blackmar, E. (1993) *The Park and the People: A History of Central Park* (Ithaca, NY: Cornell University Press), p. 330

26. Holmes, O. W. (1892) *Collected Poems,* (New York: Houghton Mifflin), vol. 1, p. 265

27. Duffy, J. (1990) *The Sanitarians: A History of American Public Health* (Urbana: University of Illinois Press)

28. Weissmann, G. (1995) The Sanitarians of Central Park. *Democracy and DNA* (New York: Hill & Wang), p. 91

29. Corinne, T. A. (2002) *Stebbins, Emma (1815–1882)* http://www.glbtq.com/arts/stebbins_e.html. Accessed Sep. 2009

30. Anonymous (1878). The Successor of Siddons. *The New York Times,* May 28, p. 10

## Quorum Sensing on the Airbus Wing

1. Gellman, B., Shulman, R. (2009) All Survive Jet's Splashdown in Hudson River: Airbus Carrying 155 People Apparently Hit Geese Minutes After Takeoff from New York. *The Washington Post,* Jan. 16, A01

2. Kropotkin, P. (1902) *Mutual Aid: A Factor of Evolution,* 1972 edition (New York: New York University Press), p. 13

3. Asad, S., Opal, S. M. (2008) Bench to Bedside Review: Quorum Sensing and the Role of Cell-to-Cell Communication During Invasive Bacterial Infection. *Crit. Care* [Epub. ahead of print] [CrossRef] [Medline] (2008) *Crit. Care* **12** (6), 236

4. Fuqua, W. C., Winans, S. C., Greenberg, E. P. (1994) Quorum Sensing in Bacteria: The LuxR-LuxI Family of Cell Density-Responsive Transcriptional Regulators. *J. Bacteriol.* **176**, 269–267 [Free Full Text]

5. Diggle, S. P., Gardner, A., West, S. A., Griffin, A. S. (2007) Evolutionary Theory of Bacterial Quorum Sensing: When Is a Signal Not a Signal? *Philos. Trans. R. Soc. Lond. B. Biol. Sci.* **362**, 1241–1249 [Abstract/Free Full Text]

6. Dwyer, J. (2009) Old Hands on the River Didn't Have to Be Told What to Do. *The New York Times,* Jan. 17, A19

7. Fuller, M. (1845) *Woman in the Nineteenth Century,* 1855 edition (Boston: J. J Jewett), p. 174

8. Collins, G. (2009) Lilly's Big Day. *The New York Times,* Jan. 28, A27

9. Leocha, C. (2009) Unsung Heroes on the Hudson—Flight Attendants on US Airways 1549, Tripso.com, Jan. 20. http://www.tripso.com/today/unsung-heroes-on-the-hudson-flight-attendants-on-us-airways-1549/. Accessed Feb. 2009

10. Wade, M. (1940) *Margaret Fuller: Whetstone of Genius* (New York: Viking Press), p. 139ff

11. Fuller, p. 253

12. Fuller, 1855 edition only, p. 390

13. Kropotkin, P. (1899) *Memoirs of a Revolutionist* (Boston and New York: Houghton, Mifflin), p. 260

14. Ibid., p. 381

15. *The Freedom Press* http://www.freedompress.org.uk/public/news.oml.html. Accessed Feb. 2009

16. Kropotkin, *Memoirs,* p. 49

17. Ford, F. M. (1971), Memories and Impressions. *The Bodley Head Ford Madox Ford,* vol. 5, (London: Bodley Head), p. 190

18. Kropotkin, *Mutual Aid,* p. 4

19. Williams, T. C., Klonowski, T. J., Berkeley, P. (1976) Angle of Canada Goose Flight Formation Measured by Radar. *The Auk* **93**, 554–559

20. McFall-Ngai, M. J., Ruby, E. G. (1991) Symbiont Recognition and Subsequent Morphogenesis as Early Events in an Animal-Bacterial Mutualism. *Science* **254**, 1491–1494 [Abstract/Free Full Text]

21. Hastings, J. W. (1971) Light to Hide By: Ventral Luminescence to Camouflage the Silhouette. *Science* **173**, 1016–1017 [Abstract/Free Full Text]

22. Jones, B. W., Nishiguchi, M. K. (2004) Counterillumination in the Hawaiian Bobtail Squid, *Euprymna Scolopes. Marine Biology* **144**, 1151–1155 [CrossRef]

23. Popat, R., Crusz, S. A., Diggle, S. P. (2008) The Social Behaviours of Bacterial Pathogens. *Brit. Med Bulletin* **87**, 63–75 [Abstract/Free Full Text]

24. Novick, R. P., Geisinger, E. (2007) Quorum Sensing in Staphylococci. *Annu. Rev. Genet.* **42**, 541–564 [CrossRef]

25. Bassler, B. L., Losick, R. (2006) Bacterially Speaking. *Cell* **125**, 237–246 [CrossRef] [Medline]

26. MacPherson, K. (2009) Bassler Wins Wiley Prize in Biomedical Sciences. *News at Princeton,* Feb. 3. http://www.princeton.edu/main/news/archive/S23/37/03S23/index.xml?section=topstories Accessed Feb. 2009

## *SiCKO* Statistics: Michael Moore and L'École de Paris

1. Morse, J. T. (1896) *Life and Letters of Oliver Wendell Holmes* (Boston and New York: The Riverside Press Cambridge, Houghton, Mifflin), p. 431

2. Cronkite, W. (1993) Quoted in review of Norwick R.-M. (2004). American Indian Health: Innovations in Health Care, Promotion, and Policy. *J. Health Care Poor and Underserved* **15**, 493–494 [CrossRef]

3. "SiCKO" Factual Backup. http://www.michaelmoore.com/sicko/checkup/. Accessed Jan. 2009

4. Bollet, A. J. (1973) Pierre Louis: The Numerical Method and the Foundation of Quantitative Medicine. *Am. J. Med. Sci.* **266**, 92–101 [CrossRef] [Medline]

5. Stein, J. (2005) Michael Moore: The Angry Filmmaker. http://www.time.com/time/subscriber/2005/time100/artists/100moore.html. Accessed Jan. 2009

6. Louis, P. C. A. (1835) *Recherches sur les effets de la saignée dans quelques maladies inflammatoires, et sur l'action de l'émétique et des vésicatoires dans la pneumonie* (Paris: Baillière)

7. World Health Organization (2000) World Health Organization Assesses the World's Health Systems http://www.who.int/whr/2000/mediacentre/pressrelease/en/index.html. Accessed Jan. 2009

8. Nolte, E., McKee, C. M. (2008) Measuring the Health of Nations: Updating an Earlier Analysis. *Health Aff.* **27**, 58–71 [Abstract/Free Full Text]

9. Dutton, P. V. (2007) France's Model Healthcare System. *Boston Globe*, Aug. 11, A11

10. Dutton, P. V. (2007) *Differential Diagnoses: A Comparative History of Health Care Problems and Solutions in the United States and France* (Ithaca, NY: Cornell University Press), p. 6

11. AERES (2008) A Review of INSERM by the International Visiting Committee: Enhancing the Future of Life Sciences and Health Research in France. http://www.inserm.fr/en/inserm/documentsstrategiques/index.html. Accessed Jan. 2009

12. Rodwin, V. J. (2005) A Comparative Analysis of Health Systems Among Wealthy Nations. In Jonas, S. Kovner, A. R. Knickman, J., eds. (2005) *Health Care Delivery in the United States* (New York: Springer), p. 175

13. Holmes, O. W. (1891) *The Autocrat of the Breakfast-Table* (Boston: Houghton Mifflin; Riverside Press), p. 172

14. Morse, p. 109

15. Ibid., p. 149

16. Ibid., p. 89

17. Osler, W. (1897) Influence of Louis on American Medicine. *Bull. Johns Hopkins Hosp.* **8**, 161–167

18. Morse, p. 180

19. Holmes, O. W. (1891) *Medical Essays* (Boston: Houghton Mifflin; Riverside Press), p. 243

20. Ibid., p. 431

21. Morse, p. 146

22. Ibid., p. 102

23. Rutkow, I. M. (1998) Anatomical Studies in Antebellum America. *Arch. Surg.* **133**, 137

24. Holmes, *Medical Essays*, p. 431; *Morse,* p. 15

### Ask Your Doctor: Justice Holmes and the Marketplace of Ideas

1. Holmes, O. W., Jr. (1919) *Abrams v. United States, 250 U.S. 616, 630 (dissent)*

2. Holmes, O. W., Jr. (1905) *In Lochner v. United States, 198 U.S. 45 (dissent)*

3. Spencer, H. (1886; orig. 1864) *The Principles of Biology* (New York: D. Appleton & Co.), p. 444

4. Tamers, S. L., Agurs-Collins, T., Dodd, K. W., Nebeling, L. (2009) U.S. and France Adult Fruit and Vegetable Consumption Patterns: An International Comparison. *Eur. J. Clin. Nutr.* **63**, 11–17 [CrossRef] [Medline]

5. Central Intelligence Agency, The World Factbook. https://www.cia.gov/library/publications/the-world-factbook /geos/fr.html

6. Organisation for Economic Co-operation and Development, Executive Summary, Health at a Glance: OECD Indicators 2005;ISBN: 92–64-01262–1. http://www.oecd.org/dataoecd/58/47/35624825.pdf

7. Hoffman, J. M., Shah, N. D., Vermeulen, L. C., Doloresco, F., Martin, P. K., Blake, S., Matusiak, L., Hunkler, R. J., Schumock, G. T. (2002) Projecting Future Drug Expenditures—2009. *Am. J. Health Syst. Pharm.* **66**, 237–257 [CrossRef]

8. Gellad, Z. F., Lyles, K. W. (2007) Direct-to-Consumer Advertising of Pharmaceuticals. *Am. J. Med.* **120**, 475–480 [CrossRef] [Medline]

9. Gagnon, M. A., Lexchin, J. (2008) The Cost of Pushing Pills: A New Estimate of Pharmaceutical Promotion Expenditures in the United States. *PLoS Med.* **5**, e1 [CrossRef] [Medline]

10. Stanley, D. (2009) Four Billion Dollars' Worth of DTC Promotion Money Goes Toward Messages Urging Consumers to "Ask Your Doctor." *Drug Topics* **153**, 44

11. Donohue, J. M., Cevasco, M., Rosenthal, M. B. (2007) A Decade of Direct-to-Consumer Advertising of Prescription Drugs. *N. Engl. J. Med.* **357**, 673–681 [Abstract/Free Full Text]

12. Brownfield, E. D., Bernhardt, J. M., Phan, J. L., Williams, M. V., Parker, R. M. (2004) Direct-to-Consumer Drug Advertisements on Network Television: An Exploration of Quantity, Frequency, and Placement. *J. Health Commun.* **9**, 491–497 [CrossRef] [Medline]

13. Khanfar, N. M., Polen, H. H., Clauson, K. A. (2009) Influence on Consumer Behavior: The Impact of Direct-to-Consumer Advertising on Medication Requests for Gastroesophageal Reflux Disease and Social Anxiety Disorder. *J. Health Commun.* **14**, 451–460 [CrossRef] [Medline]

14. Kessler, D. A., Pines, W. L. (1990) The Federal Regulation of Prescription Drug Advertising and Promotion. *JAMA* **264**, 2409–2415 [Abstract/Free Full Text]

15. Weissmann, G. (2009) The Atlanta Falcon and Tono-Bungay: Dietary Supplements as Subprime Drugs. *FASEB J.* **23**, 1279–1282 [Free Full Text]

16. *United States v. Johnson, 221 U.S. 488* 1911

17. Morse, J. T. (1896) *Life and Letters of Oliver Wendell Holmes* (Cambridge, MA: Houghton, Mifflin), pp. 202–203

18. U.S. Food and Drug Administration, Significant Dates in U.S. Food and Drug Law History. http://www.fda.gov/AboutFDA/WhatWe Do/History/Milestones/ucm128305.htm

19. Hopkins, W. W. (1996) The Supreme Court Defines the Marketplace of Ideas. *Journalism Mass Comm. Q.* **73**, 40–53

20. *Hammer v. Dagenhart, 247 U.S. 251* 1918

21. Howe, M. D. W. (1911) *Holmes-Pollock Letters* (Cambridge, MA: Harvard University Press), p. 58

22. The Darwin Centenary at Cambridge. *Science* 1909; 30, 52–53[Free Full Text]

23. *Buck v. Bell, 274 U.S. 200* 1927

## Filter the Dogs: Microbial Mishaps in Massachusetts

1. Town of Falmouth: Boil Order Information History. http://www.falmouthmass.us/deppage.php?number=417. Accessed June 18, 2010

2. Kennedy, E. P. (2010) personal communication with Gerald Weissmann, June 18

3. Riley, M., Abe, T., Arnaud, M. B., Berlyn, M. K., Blattner, F. R., Chaudhuri, R. R., Glasner, J. D., Horiuchi, T., Keseler, I. M., Kosuge, T., Mori, H., Perna, N. T., Plunkett, G., III, Rudd, K. E., Serres, M. H., Thomas, G. H., Thomson, N. R., Wishart, D., Wanner, B. L. (2006) *Escherichia coli* K–12: A Cooperatively Developed Annotation Snapshot—2005. *Nucleic Acids Res.* **34**, 1–9 [Abstract/Free Full Text]

4. Holmes, O. W. (1863) Address to the National Sanitary Association. *Amer. J. Med. Sci.* **45**, 157–160 [CrossRef]

5. Wickett, S. (2010) "Falmouth Water Must Be Boiled." *Boston Globe,* June 17, B3

6. Trails.com, "Falmouth Town Forest—Long Pond Trail." http://www.trails.com/tcatalog_trail.aspx?trailid=XMR016–007

7. Kazarian, C. (2010) Water Department Continues to Work in High Gear. *Falmouth Enterprise,* July 9, 1

8. Kazarian, C. (2010) Investigation Puts Blame for Water Problems on Two Staff Members, *Falmouth Enterprise,* July 9, 1

9. Sterling, E. J. (2009) Closing remarks, Milstein Science Symposium: American Museum of Natural History. http://symposia.cbc.amnh.org/archives/health/

10. Knowles, A. S. (2010) personal communication with Gerald Weissmann, June 20

11. Lederberg, J. (2000) Microbiology's World Wide Web, Project Syndicate 200 -12-01, http://www.project-syndicate.org/commentary/led1/English

12. Sogin, M. (2010) Op-Ed, *Falmouth Enterprise*, June 25, A4

13. Waksman Foundation for Microbiology, http://www.waksman foundation.org/

14. Tatum, E. L., Lederberg, J. (1947) Gene Recombination in the Bacterium *Escherichia coli. J. Bacteriol.* **53**, 673–684 [Free Full Text]

15. Meselson, M., Stahl, F. W. (1958) The Replication of DNA in *Escherichia coli. Proc. Natl. Acad. Sci. USA* **44**, 671–682 [Free Full Text]

16. Pardee, A. B., Williams, I. (1952) The Increase in Desoxyribonuclease of Virus-Infected *E. coli. Arch. Biochem. Biophys.* **40**, 222–223 [CrossRef] [Medline]

17. Changeux, J. P. (1960) On the Biochemical Expression of Genetic Determinants of *Escherichia coli* Introduced to *Salmonella typhimurium. C. R. Hebd. Seances Acad. Sci.* **250**, 1575–1577 [Medline]

18. Kornberg, H. L. (2003) Memoirs of a Biochemical Hod Carrier. *J. Biol. Chem.* **278**, 9993–10001 [Free Full Text]

19. Kennedy, E. P. (1992) Sailing to Byzantium. *Annu. Rev. Biochem.* **61**, 1–28 [CrossRef] [Medline]

20. Miller, K. J., Kennedy, E. P., Reinhold, V. N. (1986) Osmotic Adaptation by Gram-Negative Bacteria: Possible Role for Periplasmic Oligosaccharides. *Science* **231**, 48–51 [Abstract/Free Full Text]

21. The Ellison Medical Foundation, Joshua Lederberg, cited in http://www.ellison foundation.org/pfbs.jsp?p=107

22. NIH HMP Working Group. Peterson, J., Garges, S., Giovanni, M., McInnes, P., Wang, L., Schloss, J. A., Bonazzi, V., McEwen, J. E., Wetterstrand, K. A., Deal, C., et al. (2009) The NIH Human Microbiome Project. *Genome Res.* **19**, 2317–2323 [Abstract/Free Full Text]

23. Holmes, O. W., Jr. (Nov 21, 1927) *Compania General De Tabacos De Filipinas vs. Collector of Internal Revenue, 275 U.S. 87, 100, dissenting opinion*

24. Osler, W. (1894) Oliver Wendell Holmes. *Bull. Johns Hopkins Hospital* **42**, 85–88

25. Holmes, O. W. (1843) The Contagiousness of Puerperal Fever. *New Engl. Quart. J. Med. Surg.* **1**, 503–530

26. Holmes, O. W. (1855) *Puerperal Fever as a Private Pestilence* (Boston: Ticknor and Fields)

27. Barker, F. (1857) Puerperal Fever in New York. *New York J. Med. Coll. Sci. 3d series* **3**, 105–107, 348–355

28. Charcot, J-M. (1895) Pasteur. *The Cosmopolitan* **18**, 271–278

29. NationMaster.com, Maternal Mortality (Most Recent) by Country, http://www.nationmaster.com/graph/hea_mat_mor-health-maternal-mortality

30. NationMaster.com, Water Contact Diseases (Most Recent) by Country, http://www.nationmaster.com/graph/hea_maj_inf_dis_wat_con_dis-infectious-diseases-water-contact-disease

## Pattern Recognition and Gestalt Psychology: The Day Nüsslein-Volhard Shouted *"Toll!"*

### *Acknowledgments*

With heartfelt thanks to Prof. Dr. Christiane Nüsslein-Volhard (Max-Planck-Institut für Entwicklungsbiologie, Tübingen, Germany) for her historic contributions—and her historic note to *The FASEB Journal*. Thanks also to Ann Weissmann (M.A., New School), who first connected the dots of pattern recognition in psychology and immunology.

### *References*

1. Willemsen, R. (2009) *Ich hatte immer wieder Heureka-Erlebnisse* (interview with Christiane Nüsslein-Volhard) Zeitonline, Sep. 4. http://www.zeit.de/2009/16/Willemsen-Nuesslein-Volhard-16. Accessed May 2010

2. Wertheimer, M. (1924) *Gestalt Theory (Address to the Kant Society)* Hayes Barton Press. http://store.vitalsource.com/product/show/L99974225?partner=vstgbs. Accessed May 2010

3. Kandel, E. (2006) *In Search of Memory* (New York: W. W. Norton), pp. 301–302

4. Takeuchi, O., Akira, S. (2010) Pattern Recognition Receptors and Inflammation. *Cell* **140**, 805–820 [CrossRef] [Medline]

5. Pajarinen, J., Mackiewicz, Z., Pöllänen, R., Takagi, M., Epstein, N. J., Ma, T., Goodman, S. B., Konttinen, Y. T. (2010) Titanium Particles Modulate Expression of Toll-like Receptor Proteins. *J. Biomed. Mater. Res. A.* **92**, 1528–1537 [Medline]

6. Maier, K. L., Alessandrini, F., Beck-Speier, I., Hofer, T. P., Diabaté, S., Bitterle, E., Stöger, T., Jakob, T., Behrendt, H., Horsch, M., Beckers, J., Ziesenis, A., Hültner, L., Frankenberger, M., Krauss-Etschmann, S., Schulz, H. (2008) Health Effects of Ambient

Particulate Matter—Biological Mechanisms and Inflammatory Responses to In Vitro and In Vivo Particle Exposures. *Inhal. Toxicol.* **20**, 319–337 [CrossRef] [Medline]

7. Rom, W. N., Reibman, J., Rogers, L., Weiden, M. D., Oppenheimer, B., Berger, K., Goldring, R., Harrison, D., Prezant, D. (2010) Emerging Exposures and Respiratory Health: World Trade Center Dust. *Proc. Am. Thorac. Soc.* **7**, 142–145 [Abstract/Free Full Text]

8. Cirl, C., Wieser, A., Yadav, M., Duerr, S., Schubert, S., Fischer, H., Stappert, D., Wantia, N., Rodriguez, N., Wagner, H., Svanborg, C., Miethke, T. (2008) Subversion of Toll-Like Receptor Signaling by a Unique Family of Bacterial Toll/Interleukin-1 Receptor Domain-Containing Proteins. *Nat. Med.* **14**, 399–406 [CrossRef] [Medline]

9. Wiens, M., Korzhev, M., Perovic-Ottstadt, S., Luthringer, B., Brandt, D., Klein, S., Müller, W. E. (2007) Toll-Like Receptors Are Part of the Innate Immune Defense System of Sponges (Demospongiae: Porifera). *Mol. Biol. Evol.* **24**, 792–804 [Abstract/Free Full Text]

10. Medzhitov, R., Preston-Hurlburt, P., Janeway, C. A., Jr. (1997) A Human Homologue of the Drosophila Toll Protein Signals Activation of Adaptive Immunity. *Nature* **388**, 394–397 [CrossRef] [Medline]

11. Anderson, K. V., Jürgens, G., Nüsslein-Volhard, C. (1985) Establishment of Dorsal-Ventral Polarity in the Drosophila Embryo: Genetic Studies on the Role of the Toll Gene Product. *Cell* **42**, 779–789 [CrossRef] [Medline]

12. Anderson, K. V., Bokla, L., Nüsslein-Volhard, C. (1985) Establishment of Dorsal-Ventral Polarity in the Drosophila Embryo: The Induction of Polarity by the Toll Gene Product. *Cell* **42**, 791–798 [CrossRef] [Medline]

13. Nüsslein-Volhard, C., Wieschaus, E. (1980) Mutations Affecting Segment Number and Polarity in Drosophila. *Nature* **287**, 795–801 [CrossRef] [Medline]

14. Nüsslein-Volhard, C. (1995) Nobel Lecture: The Identification of Genes Controlling Development in Flies and Fishes. Ringertz, N., eds. *Nobel Lectures, Physiology or Medicine 1991–1995*, (1997) (Singapore: World Scientific Publishing Co.) http://nobelprize.org/nobel_prizes/medicine/laureates/1995/nusslein-volhard-lecture.html. Accessed May 2010

15. Wieschaus, E. F. (1995) Nobel Lecture: From Molecular Patterns to Morphogenesis: The Lessons from Drosophila. Ringertz, N. ed. *Nobel Lectures.* http://nobelprize.org/nobel_prizes/medicine/laureates/1995/wieschaus-lecture.html. Accessed May 2010

16. Hashimoto, C., Hudson, K. L., Anderson, K. V. (1988) The Toll Gene of Drosophila, Required for Dorsal-Ventral Embryonic Polarity, Appears to Encode a Transmembrane Protein. *Cell* **52**, 269–279 [CrossRef] [Medline]

17. Rock, F. L., Hardiman, G., Timans, J. C., Kastelein, R. A., Bazan, J. F. (1998) A Family of Human Receptors Structurally Related to Drosophila Toll. *Proc. Natl. Acad. Sci. U S A* **95**, 588–593 [Abstract/Free Full Text]

18. Beutler, B. A. (2009) TLRs and Innate Immunity. *Blood* **113**, 1399–1407 [Abstract/Free Full Text]

19. Schnare, M., Holt, A. C., Takeda, K., Akira, S., Medzhitov, R. (2000) Recognition of CpG DNA Is Mediated by Signaling Pathways Dependent on the Adaptor Protein MyD88. *Curr. Biol.* **10**, 1139–1142 [CrossRef] [Medline]

20. Goldman, R. D. (1971) The Role of Three Cytoplasmic Fibers in BHK-21 Cell Motility. Microtubules and the Effects of Colchicine. *J. Cell Biol.* **51**, 3752–3762

21. Hoffstein, S., Weissmann, G. (1978) Microfilaments and Microtubules in Calcium Ionophore-Induced Secretion of Lysosomal Enzymes from Human Polymorphonuclear Leukocytes. *J. Cell Biol.* **78**, 769–781 [Abstract/Free Full Text]

22. Ueda, M., Graf, R., MacWilliams, H. K., Schliwa, M., Euteneuer, U. (1997) Centrosome Positioning and Directionality of Cell Movements. *Proc. Natl. Acad. Sci. USA* **94**, 9674–9678 [Abstract/Free Full Text]

23. King, D. B., Wertheimer, M. (2005) *Max Wertheimer and Gestalt Theory* (New Brunswick, NJ: Transaction)

24. Goethe, J. W. (1806) *Metamorphose der Tiere* [quoted in German, Ref. 16. (Weissmann, G., trans)]

25. King, Wertheimer, p. 98

26. Wertheimer, M. (1923) Untersuchungen zur Lehre von der Gestalt II. *Psycologische Forschung* **4**, 301–350 [CrossRef]

27. Behrens, R. R. (1998) Art, Design and Gestalt Theory. *Leonardo* **31**, 299–303 [CrossRef]

28. Wertheimer, M (1912) Experimentelle Studien uber das Sehen von Bewegung. *Z. Psychologie* **61**, 161–265

29. Köhler, W. (1926) *The Mentality of Apes* (New York: Harcourt, Brace)

30. Sokal, M. M. (1984) The Gestalt Psychologists in Behaviorist America. *Am. Hist. Review* **89**, 1240–1263 [CrossRef]

31. Köhler, W. (1959) Gestalt Psychology Today. *Am. Psychologist* **14**, 727–734 [CrossRef]

### Not by the Sword, but Disease: Doctor Howe and General Shinseki

1. Richards, L. E. ed. (1909) *Letters and Journals of Samuel Gridley Howe* (Boston: D. Estes), p. 2

2. Shinseki, E. K. As quoted in: Deputy Secretary Wolfowitz Interview with *The New Yorker* (June 18, 2002) http://www.defenselink.mil/transcripts/transcript.aspx?transcriptid=3527. Accessed Dec. 2008

3. Agence France Presse (2008) Obama Vows to Improve Treatment of Vets, Dec. 8. http://www.afp/article/ALeqM5hnGfBeeXOnZn 9PCMx1gLDhyf5pJQ. Accessed Dec. 2008

4. United Press International (2007) Bush Attends Ceremonial Swearing In of Veterans Affairs Secretary James Peake, Dec. 20 http://www.upi.com/topic/JamesPeake/. Accessed Dec. 2008

5. Sanger, D. E. (2004) Vatican Envoy to Head Veterans Affairs. *The New York Times*, Dec. 10, 18

6. Associated Press (2007) Bush VA Secretary Abruptly Steps Down. *Charleston Gazette*, July 18, p. 5D

7. Yen, H. (2007) Injured Iraq War Veterans Sue Bush's VA Head for Poor Care, Cheating, Associated Press, July 23

8. Anonymous (2007) President George W. Bush Delivers Remarks at the Veterans Affairs Medical Center, Political Transcript Wire, Aug. 13. ABI/INFORM Trade & Industry database. Document ID: 1319160631. Retrieved Dec. 12, 2008

9. VA Hospitals: Issues and Challenges for the Future, 1998 United States General Accounting Office http://www.gao.gov/cgi-bin/getrpt?HEHS-98–32. Accessed Dec. 2008

10. NIH Funding by Institute, 2006 American Association for the Advancement of Science. http://www.aaas.org/spp/rd/health07p.pdf. Accessed Dec. 2008

11. Kirk, M. (director, writer) (2004) *Frontline: Rumsfeld's War* [television broadcast transcript] PBS http://www.pbs.org/wgbh/pages/frontline/shows/pentagon/etc/script.html. Accessed Dec. 2008

12. Fallows, J. (2004) Blind into Baghdad. *The Atlantic* **293**, 52–57

13. Gordon, M. R., Mazzetti, M. (2006) Top U.S. General Warns Against Troop Withdrawal from Iraq. *The New York Times: International Herald Tribune,* Nov. 16. http://www.iht.com/articles/2006/11/16/news/policy.php. Accessed Dec. 2008

14. *Who's Who of Asian Americans: General Eric K. Shinseki,* AsianAmerican.Net. http://www.asianamerican.net/bios/GeneralShinseki.html. Accessed Dec. 2008

15. Stillé, C. J. (1866) *History of the United States Sanitary Commission, Being the General Report of Its Work During the War of the Rebellion* (Philadelphia: J. B. Lippincott), p. 8

16. Goodwin, D. K. (2005) *Team of Rivals* (New York: Farrar, Straus & Giroux), p. 703

17. Meltzer, M. A. (1964) *Light in the Dark: The Life of Samuel Gridley Howe* (New York: Crowell)

18. Howe, S. G. (1828) *An Historical Sketch of the Greek Revolution* (New York: White, Gallaher & White)

19. Richards, p. 1

20. Ibid.

## Science as Oath and Testimony: Joshua Lederberg

1. Lederberg, J. (1978) Robin, E. D. ed. *Claude Bernard and the Internal Environment* (Stanford, CA: Stanford University Press), p. 271

2. Lederberg, J. (1991) Communication as the Root of Scientific Progress (presentation at 1991 Woods Hole Conference of International Scientific Editors). Stefik, M. eds. (1993) *Internet Dreams: Archetypes. Myths and Metaphors* (MIT Press Cambridge, MA: MIT Press), p. 41

3. Lederberg (1978), p. 272

4. Roundtable: A Meeting of Biological and Philosophical Minds. *The New York Times,* Mar. 1983, A8

5. *Profiles in Science: The Joshua Lederberg Papers* http://profiles.nlm.nih.gov/BB/. Accessed Aug. 2008

6. Ryan, F. J., Lederberg, J. (1946) Reverse-Mutation and Adaptation in Leucineless Neurospora. *Proc. Natl. Acad. Sci. U S A.* **32**, 163–173 [Free Full Text]

7. Lederberg, J. (1946) A Nutritional Concept of Cancer. *Science* **104**, 428 [Free Full Text]

8. Avery, O. T., MacLeod, C. M., McCarty, M. (1944) Studies on the Chemical Nature of the Substance Inducing Transformation of Pneumococcal Types. Induction of Transformation by a Desoxyribonucleic Acid Fraction Isolated from Pneumococcus Type III. *J. Exp. Med.* **79**, 137–158 [Abstract]

9. Tatum, E. L., Lederberg, J. (1947) Gene Recombination in the Bacterium *Escherichia coli. J. Bacteriol.* **5**, 673–684

10. Lederberg, J. (1996) Genetic Recombination in *Escherichia coli*: Disputation at Cold Spring Harbor, 1946–1996. *Genetics* **144**, 439–443 [Medline]

11. Brink, R. A. (1974) Early History of Genetics at the University of Wisconsin-Madison. http://profiles.nlm.nih.gov/BB/G/K/X/V/. Accessed Aug. 2008

12. Lederberg, E. M., Lederberg, J. (1953) Genetic Studies of Lysogenicity in *Escherichia Coli*. *Genetics* **38**, 51–64 [Free Full Text]

13. Zinder, N. D., Lederberg, J. (1952) Genetic Exchange in Salmonella. *J. Bacteriol.* **64**, 679–699 [Free Full Text]

14. Lederberg, J. (1958) Speech at the Nobel Banquet in Stockholm. http://nobelprize.org/nobelprizes/medicine/laureates/1958/lederberg-speech.html. Accessed Aug. 2008

15. Lederberg, J. J. (1999) J. B. S. Haldane (1949) on Infectious Disease and Evolution. *Genetics* **153**, 1–3[Free Full Text]

16. Lederberg, J. (1969) The Agony of the Scientists. *The Washington Post*, Sep. 20, p. 21

17. Lederberg, J. (1966) Using Bigotry Against Bias. *The Washington Post*, July 31, p. 28

18. Pew Scholars Program in the Biomedical Sciences: Program Description http://www.futurehealth.ucsf.edu/biomed/scholdes.html. Accessed Aug. 2008

19. Ellison Medical Foundation, http://www.ellison-med-fn.org/. Accessed Aug. 2008

20. Profiles in Science: The Joshua Lederberg Papers. Letter from Joshua Lederberg to Eugene Garfield, May 9, 1959. http://profiles.nlm.nih.gov/BB/A/I/X/L/. Accessed Aug. 2008

21. Hirsch, J. E. (2005) An Index to Quantify an Individual's Scientific Research Output. *Proc. Natl. Acad. Sci. U S A.* **102**, 16569–16572 [Abstract/Free Full Text]

22. Lederberg, J. (Mar. 2007) personal communication

## X-Ray Politics: The Nazi War on Röntgen and Einstein

1. Dam, H. J. W. (1896) The New Marvel in Photography. *McClure's Magazine* **5**, 413 (Munich: J. F. Lehmann)

2. Lenard, P. (1936) *Deutsche Physik* **1**, ix

3. Philips, K. (2010) *Snowcapping the Week's End. The Caucus: The Politics and Government blog of The New York Times*, Feb. 5

4. Anonymous (1933) Nationality of Jews; Locker-Lamson's Bill. *Times of London,* July 27, p. 7

5. *Judgment at Nuremberg* (1961), United Artists film; Stanley Kramer, director; Abby Mann, writer

6. Etter, L. E. (1946) Some Historical Data Relating to the Discovery of the Roentgen Rays. *Am. J. Roent. Radium Ther.* **56**, 220–231

7. Friedman, R. M. (2001) *The Politics of Excellence: Behind the Nobel Prize in Science* (New York: Times Books), p. 120

8. Wilhelm Conrad Röntgen The Nobel Prize in Physics 1901. *Nobel Lectures, Physics 1901–1921* (1967) (Amsterdam: Elsevier Publishing Company) http://nobelprize.org/nobel_prizes/physics/laureates/1901/rontgen-bio.html. Accessed Apr. 2010

9. Jerome, F., Taylor, R. (2006) *Einstein on Race and Racism* (New Brunswick, NJ: Rutgers University Press), pp. 3–11

10. Israel, H., Ruckhaber, E., Weinmann, R. (1931) *Hundert Autoren Gegen Einstein* (Leipzig: R. Voigtlander)

11. Albert Einstein Anecdotes, The Gold Scales, http://oaks.nvg.org/sa5ra17.html# einstein-anecdotes. Accessed Apr. 2010

12. Brian, D. (1996) *Einstein: A Life* (New York: John Wiley), p. 250

13. Weinrich, M. (1946) *Hitler's Professors: The Part of Scholarship in Germany's Crimes Against the Jewish People*, new edition 1991 (New Haven: Yale University Press), p. 11

14. Famous World Trials: Nuremberg Trials 1945–1949. Defendants in the Major War Figures Trial: Rosenberg http://www.law.umkc.edu/faculty/projects/ftrials/nuremberg/meetthedefendants.html. Accessed Apr. 2010

15. Glasser, O. (1958) *Dr. W. C. Roentgen,* 2nd ed. (Springfield, IL: Thomas)

16. Institut Curie, Chronologies: Milestones in the History of Radioactivity. http://www.curie.fr/fondation/musee/chronologies.cfm/lang/_gb.htm. Accessed Apr. 2010

17. Roentgen, W. C. (1895) Ueber eine Neue Art von Strahlen. *Sitzber Physik Med Ges Würzburg,* 132–141

18. Anonymous (1896) Roentgen's Discovery. *New York Tribune,* Jan. 31, 14

19. Anonymous (1896) Exploring the Radiance. *New York Tribune,* Feb. 9, 12

20. Pupin, M. L. (1996) Roentgen Rays. *Science* **3**, 231

21. Frost, E. D. (1896) Experiments on the X-rays. *Science* **3**, 235 [Free Full Text]

22. Fujikawa, A., Tamura, T., Naoi, Y., Kimura, M., Kominami, M., Hayashi, T. (2003) Scenes from the Past. The Dawn of Radiology in Japan. *Radiographics* **23**, 1011–1017 [Free Full Text]

23. Oestreich, A. E. (1996) Centennial History of African-American Radiology. *Am. J. Radiol.* **166**, 255–258 [Abstract/Free Full Text]

24. Kogelnik, H. D. (1997) Inauguration of Radiotherapy as a New Scientific Speciality by Leopold Freund 100 Years Ago. *Radiother. Oncol.* **42**, 203–211 [CrossRef] [Medline]

25. Weissmann, G. (1990) *The Doctor with Two Heads and Other Essays* (New York: Alfred A. Knopf), p. 3ff

26. Erdös, E. G. (2006) The ACE and I: How ACE Inhibitors Came to Be. *FASEB J.* **20**, 1034–1038 [Free Full Text]

**Wild Horses and *The Doctor's Dilemma***

1. Associated Press (2009) OSU Rejects Anthrax Study on Baboons, Nov. 30

2. Shaw, G. B. (2004; orig. 1909) Preface to *The Doctor's Dilemma. Pygmalion and Three Other Plays* (New York: Barnes and Noble Classics), p. 93

3. Simpson, S. (2009) Anthrax Study Rejected by OSU: Euthanasia of Primates May Be to Blame for Decision to Cancel Veterinary School Project. *Oklahoman,* Nov. 30, 1

4. Akst, J. (2009) School Halts Baboon Anthrax Study.*The Scientist.com,* Dec. 1, http://www.the-scientist.com/blog/display/56193/

5. Stearns-Kurosawa, D. J., Lupu, F., Taylor, F. B., Jr, Kinasewitz, G., Kurosawa, S. (2006) Sepsis and Pathophysiology of Anthrax in a Nonhuman Primate Model. *Am. J. Pathol.* **169**, 433–444 [Abstract/Free Full Text]

6. Taylor, F. B., Jr., Chang, A., Esmon, C. T., D'Angelo, A., Vigano-D'Angelo, S., Blick, K. E. (1987) Protein C Prevents the Coagulopathic and Lethal Effects of *Escherichia coli* Infusion in the Baboon. *J. Clin. Invest.* **79**, 918–925 [Medline]

7. Oklahoma State U. halts an anthrax study, citing plan to euthanize baboons. (Nov. 30, 2009) *The Chronicle of Higher Education.* The Ticker, http://chronicle.com/blogs/ticker/oklahoma-state-u-halts-an-anthrax-study-citing-plan-to-euthanize-baboons/9016; blog posted Dec. 1, 2009.

8. American Veterinary Medical Association (2009) Donor Doesn't Want Money Going to Veterinary College: Cites Allegations of Inhumane Treatment of Animals. *JAVMA News. College News,* Apr. 1. http://www.avma.org/onlnews/javma/apr09/090401s.asp

9. O'Toole, A. (2009) Horse Sensitivity: Show at OSU Pushes Preserve for Mustangs. *Tulsa World,* Nov. 20, A15

10. OKState.com, Cowboy Football, 2009 Schedule. http://www.okstate.com/sports/m-footbl/sched/okst-m-footbl-sched.html

11. Dunham, J. (2009) Former Owner of Racehorses Now Works to Save Them. *The New York Times,* May 1, C2

12. Help Save America's Wild Horses, http://www.madeleinepickens.com/

13. Howard, B. (Dec. 1, 2009) Kudos for a Great Decision!—OSU President Cancels Anthrax Study Proposal Requiring Primate Euthanasia. *DVM Newsmagazine,* http://www.madeleinepickens.com/news/osu-president-cancels-antrax-study-proposal-requiring-primate-euthanasia/

14. FASEB (2009) FASEB Statement on the Cancellation of OSU Primate Anthrax Project. *FASEB News,* Dec. 2. http://www.faseb.org/pdf/OSU_anthrax_12_2_09.pdf

15. Schiller, J. (1967) Claude Bernard and Vivisection. *J. Hist. Med. Allied Sci.* **22**, 246–260 [Medline]

16. Foster, M. (1899) *Claude Bernard* (Master of Medicine Series) (London: T. Fisher Unwin), pp. 163–164

17. Bernard, C. (1974) *Lettres à Madame R* (Letters to Madame Raffalovich sent from St. Julien between 1869 and 1878; introduction by Jacqueline Sonolet) (France: l'Imprimerie de France), p. 185

18. Cardon, P. (1993) A Homosexual Militant at the Beginning of the Century: Marc André Raffalovich. *J. Homosex.* **25**, 183–191 [CrossRef] [Medline]

19. Raffalovich, M. A. (1905) À propos du syndicat des uranistes. *Arch. Anth. Crim.* **20**, 283–286

20. Shaw, G. B. (2008; orig. 1909) *The Doctor's Dilemma* (New York: New Vision)

21. Regan, T. (Apr. 22, 1988) A Declaration of War on Vivisection. Abolitionist-Online. http://www.abolitionist-online.com/article-issue 03_decclaration.war_tom.regan.shtml

22. Shaw, *The Doctor's Dilemma*, p. 33

## Glass Ceilings at the Nobel Prizes

1. Quinn, S. (1995) *Marie Curie: A Life* (New York: Simon & Schuster), p. 327

2. Bombardieri, M. (2005) Summers' Remarks on Women Draw Fire, *The Boston Globe,* Jan. 17

3. Mundy, L. (2009) Success Is in Her DNA: Greider Wins the Nobel Prize in a Field in Which Relatively Few Women Thrive, *The Washington Post,* Oct. 20

4. Smith, A. Interviewer (2009) Elizabeth H. Blackburn: The Nobel Prize in Physiology or Medicine 2009, Oct. 5, Nobelprize.org. http://nobelprize.org/nobel_prizes/medicine/laureates/2009/blackburn-telephone.html

5. Nobelprize.org. Meet the Nobel laureates! http://nobelprize.org

6. Szostak, J. W., Blackburn, E. H. (1982) Cloning Yeast Telomeres on Linear Plasmid Vectors. *Cell* **29**, 245–255 [CrossRef] [Medline]

7. Greider, C. W., Blackburn, E. H. (1985) Identification of a Specific Telomere Terminal Transferase Activity in Tetrahymena Extracts. *Cell* **43**, 405–413 [CrossRef] [Medline]

8. Greider, C. W., Blackburn, E. H. (1989) A Telomeric Sequence in the RNA of Tetrahymena Telomerase Required for Telomere Repeat Synthesis. *Nature* **337**, 331–337 [CrossRef] [Medline]

9. Smith, A. Interviewer (2009) Ada E. Yonath: The Nobel Prize in Chemistry 2009, Oct. 7, Nobelprize.org. http://nobelprize.org/nobel_prizes/chemistry/laureates/2009/yonath-telephone.html

10. Bartels, H., Gluehmann, M., Hansen, H., Auerbach, T., Franceschi, F., Yonath, A. (2001) High-Resolution Structures of Ribosomal Subunits: Initiation, Inhibition, and Conformational Variability. *Cold Spring Harb. Symp. Quant. Biol.* **66**, 43–56 [Abstract/Free Full Text]

11. Philips, W. (1968; orig. 1884) *Woman's Rights. Speeches, Lectures and Letters* (New York: Negro Universities Press), p. 16

12. Fröman, N. (1996) Marie and Pierre Curie and the Discovery of Polonium and Radium, Dec. 1, Nobelprize.org. http://nobelprize.org/nobel_prizes/physics/articles/curie/index.html

13. Nobelprize.org. The Nobel Prize in Chemistry 1911. Presentation Speech. http://nobelprize.org/nobel_prizes/chemistry/laureates/1911/press.html

14. Nobelprize.org. Marie Curie. The Nobel Prize in Chemistry 1911. http://nobelprize.org/nobel_prizes/chemistry/laureates/1911/marie-curie-lecture.html

15. Quinn, p. 231

16. Ibid. p. 154

17. Ibid. p. 295

18. Ibid. p. 316

19. Ibid. p. 290

20. Ibid. p. 280

21. Lowen, L. Work Life Balance—Nobel Winner Elizabeth Blackburn's Work Life Balance Story. http://womensissues.about.com/od/intheworkplace/a/WorkLifeBalanceElizabeth Blackburn.htm.

**Medea and the Microtubule**

1. Re-engineering the Clinical Research Enterprise (2009) NIH Roadmap for Medical Research, http://nihroadmap.nih.gov/clinicalresearch/overview-translational.asp. Accessed July 2009

2. Charen, T. (1951) The Etymology of Medicine. *Bull. Med. Libr. Assoc.* **39**, 216–221 [Medline]

3. Euripides (431 BCE) Svarlien, D. A. trans. *Medea* (2008) (Indianapolis: Hackett), p. 18

4. Scott, A. O. (2009) Review of "Tyler Perry's Madea Goes to Jail." *The New York Times*, Feb. 21, C1

5. Science Translational Medicine: Integrating Medicine and Science 2009. http://www.sciencemag.org/marketing/stm/. Accessed July 2009

6. Lee, W.-H., Languino, L. R., Hung, M.-C., Dubinett, S. M., Iczkowski, K. A., Wang, D. (2009) Editorial: The Launch of the *American Journal of Translational Research. Am. J. Transl. Res.* **1**,1

7. Marincola, M. J. (2003) Editorial: Translational Medicine: A Two-Way Road. *J. Transl. Med.* **1**, 1 [CrossRef] [Medline]

8. Laurence, J. (2006) Editorial: Translating Translational Research. *Translational Research* **148**, 1 [CrossRef] [Medline]

9. Andrews, N., Burris, J. E., Cech, T. R., Coller, B. S., Crowley, W. F., Jr., Gallin, E. K., Kelner, K. L., Kirch, D. G., Leshner, A. I., Morris, C. D., Nguyen, F. T., Oates, J., Sung, N. S. (2009) Translational Careers. *Science* **324**, 85 [Abstract/Free Full Text]

10. Anonymous (2000) Genetic Secret Unlocked: Breakthrough Could Change Health Care. *Daily Gleaner*, June 27, 1, Fredericton NB, Canada

11. Dennis, E. A. (2009) Lipidomics Joins the Omics Evolution. *Proc. Natl. Acad Sci. USA* **106**, 2089–2090 [Free Full Text]

12. FANTOM ConsortiumSuzuki, H., Forrest, A. R., van Nimwegen, E., Daub, C. O., Balwierz, P. J., Irvine, K. M., Lassmann, T., Ravasi, T., Hasegawa, Y., de Hoon, M. J., et al. (2009) The Transcriptional Network That Controls Growth Arrest and Differentiation in a Human Myeloid Leukemia Cell Line. *Nat. Genet.* **41**, 553–562 [CrossRef] [Medline]

13. Pauling, L., Itano, H. A., Singer, S. J., Wells, I. C. (1949) Sickle Cell Anemia: A Molecular Disease. *Science* **110**, 543–558 [Free Full Text]

14. Borges, J. L. (1943) quoted in Waisman, S. ed. *Borges and Translation: The Irreverence of the Periphery* (Lewisburg, PA: Bucknell University Press), p. 113

15. Seneca (1st century BCE) (1986) *Medea*, Ahl, F. trans. (Ithaca, NY: Cornell University Press), p. 84

16. Eigsti, O. J., Dustin, P. (1955) *Colchicine* (Ames: Iowa State College Press), p. 3ff

17. Gayley, C. M., Bulfinch, T. (1893) *The Classic Myths in English Literature* (Boston and New York: Ginn and Company), p. 244ff

18. Porter, R., Rousseau, G. S. (1998) *Gout: The Patrician Malady* (New Haven: Yale University Press)

19. Weissmann, G. (2007) *Galileo's Gout* (New York: Bellevue Literary Press), p. 13ff

20. Duckworth, D. (1889) *A Treatise on Gout*, (Philadelphia: Blakiston), p. 348

21. Ahern, M. J., Reid, C., Gordon, T. P., McCredie, M., Brooks, P. M., Jones, M. (1987) Does Colchicine Work? The Results of the First Controlled Study in Acute Gout. *Aust. NZ. J. Med.* **17**, 301–304 [Medline]

22. Pernice, B. (1889) Sulla cariocinesi delle cellule epiteliale e dell endotelio dei vasi della mucosa dello stomato e dell' intestino, nello studio gastroenterite sperimentale (nell avvelenamento per colcico). *Sicilia Med.* **1**, 265–279

23. Dustin, A. (1934) Contributions à l'étude des poisons caryoclastiques sur les tumeurs animals. *Bull. Acad. Roy. Méd. Belg.* **11**, 187–502

24. Eigsti, O. J. (1938) A Cytological Study of Colchicine Effects in the Induction of Polyploidy in Plants. *Proc Natl Acad Sci USA* **24**, 56–63 [Free Full Text]

25. Levine, M. (1945) Colchicine and X-Rays in the Treatment of Plant and Animal Overgrowths. *Botanical Review* **11**, 145–180 [CrossRef]

26. Borisy, G. G., Taylor, E. W. (1967) The Mechanism of Action of Colchicine. Binding of Colchicine-3H to Cellular Protein. *J. Cell Biol.* **34**, 525–533 [Abstract/Free Full Text]

27. Borisy, G. G., Taylor, E. W. (1967) The Mechanism of Action of Colchicine. Colchicine Binding to Sea Urchin Eggs and the Mitotic Apparatus. *J. Cell Biol.* **34**, 535–548 [Abstract/Free Full Text]

28. Mohri, H. (1968) Amino-Acid Composition of "Tubulin" Constituting Microtubules of Sperm Flagella. *Nature* **217**, 1053–1054 [CrossRef] [Medline]

29. Kulic, I. M., Brown, A. E., Kim, H., Kural, C., Blehm, B., Selvin, P. R., Nelson, P. C., Gelfand, V. I. (2008) The Role of Microtubule Movement in Bidirectional Organelle Transport. *Proc. Natl. Acad. Sci. USA* **105**, 10011–10016 [Abstract/Free Full Text]

30. Zurier, R. B., Hoffstein, S., Weissmann, G. (1973) Mechanisms of Lysosomal Enzyme Release from Leucocytes. I. Effect of cyclic nucleotides and colchicine. *J. Cell Biol.* **58**, 27–41 [Abstract/Free Full Text]

31. McCarty, D. J. (2008) Urate Crystals, Inflammation, and Colchicine. *Arthritis Rheum.* **58**, S20–S24 [CrossRef] [Medline]

32. Gill, R. K., Borthakur, A., Hodges, K., Turner, J. R., Clayburgh, D. R., Saksena, S., Zaheer, A., Ramaswamy, K., Hecht, G., Dudeja, P. K. (2007) Mechanism Underlying Inhibition of Intestinal Apical Cl/OH Exchange Following Infection with Enteropathogenic. *E. coli. J. Clin. Invest.* **117**, 428–437

## Wiki-Science and Molière's Beast

1. Molière, J-B. P. *Tartuffe* and *The Misanthrope* (2009), Steiner, P. L., trans. (Indianapolis/Cambridge: Hackett). pp. 158ff

2. Unsolicited e-mail from WebmedCentral (Sep. 17, 2010) to GW [support@webmedcentral.com]

3. Sweet, J. (2010) WebmedCentral and Ginger Pee (blog comment). http://pleion.blogspot.com/2010/09/webmedcentral-and-ginger-pee.html. Accessed Dec. 2010

4. Hirson, D. (1992) *La Bête* (New York: Dramatists Play Service), p. 122

5. Johnson, R. (2010) From Facebook to WikiLeaks: The Good, Bad and Ugly of Technology. *Los Angeles Times,* Dec. 19 http://articles.latimes.com/2010/dec/19/entertainment/la-ca-social-20101219. Accessed Dec. 2010

6. http://www.ncbi.nlm.nih.gov/pmc/. Accessed Dec. 2010

7. http://www.webmd.com. Accessed Dec. 2010

8. http://www.webmedcentral.com. Accessed Dec. 2010

9. Jaseja, H. (2010) Shoe-Smell Application as a First Aid Interventional Measure in Controlling Epileptic Attacks in an Urban Population in India: A Fortuitous Empirical Finding. WebmedCentral NEUROLOGY 2010;1(9):WMC00791. Accessed Dec. 2010

10. Briggs, L. (2010) Fibromyalgia and Vehicular Trauma: A Case Report. WebmedCentral ALTERNATIVE MEDICINE 2010,1(10) WMC001001. Accessed Dec. 2010

11. Rodriguez-Fuentes, G., Luna-Ramírez, K., Soto, M. (2010) Sunscreen Use Behaviour and Most Frequently Used Active Ingredients Among Beachgoers on Cancun, Mexico. WebmedCentral DERMATOLOGY 2010;1(12):WMC001364. Accessed Dec. 2010

12. Tan, U. (2010) Two New Cases of Uner Tan Syndrome: One Man with Transition from Quadrupedalism to Bipedalism; One Man with Consistent Quadrupedalism. WebmedCentral NEUROLOGY 2010;1(9):WMC00645. Accessed Dec. 2010

13. http://www.plos.org. Accessed Dec. 2010

14. http://www.webmedcentral.com/reviewers. Accessed Dec. 2010

15. Hubbard, E. (1923) *Roycroft Dictionary and Book of Epigrams* (East Aurora, NY: Roycrofters Press), p. 35

16. Cunningham, W. (2003) Correspondence on the Etymology of Wiki. http://c2.com/doc/etymology.html. Accessed Dec. 2010

17. Smith, R. (2010) Post-publication Review. Embrace the "Market of Ideas." British Medical Journal 341, c5148

18. Young, N. S., Ioannidis, J. P. A., Al-Ubaydli, O. (2008) Why Current Publication Practices May Distort Science. *PLoS Med* 5, e201.oi:10.1371/journal.pmed.0050201

19. Scott, V. (2000) *Molière: A Theatrical Life* (Cambridge: Cambridge University Press), p. 240

20. Park, K., Daston, L. (2006) *The Cambridge History of Science; Vol. 3, Early Modern Science* (New York: Cambridge University Press), p. 160

### Arts and Science: Lewis Thomas and F. Scott Fitzgerald

1. Fitzgerald, F. S. (1996; orig. 1934) *Tender Is the Night* (New York: Scribner), p. 119ff

2. Thomas, L. (1979) How to Fix the Premedical Curriculum. *Medusa and the Snail* (New York: Viking), p. 137

3. Tilghman, S. M. (2010) *The Future of Science Education in the Liberal Arts College,* Jan. 5, Princeton University. http://www.princeton.edu/president/speeches/20100105/

4. Weissmann, G. (2004) Lewis Thomas (1913–1993). *Biographical Memoirs,* Washington, DC, National Academy of Sciences, 85, 3–23

5. Smith, Z. (2003) *The Autograph Man* (London: Penguin), p. 19

6. Sawhill, J. C. (1979) The Unlettered University. *Mod. Lang. J.* **63**, 281–285

7. Douthat, R. (2010) What Killed the Liberal Arts? The Opinion Pages *The New York Times,* Mar. 19. http://douthat.blogs.nytimes.com/2010/03/19/what-killed-the-liberal-arts/

8. Connor, W. R., Ching, C. (2010) Liberal Arts I: They Keep Chugging Along. *Inside Higher Ed,* Oct. 1. http://www.insidehighered.com/views/2010/10/01/connor

9. Wilson, E. (1975) *The Twenties* (Edel, L., ed.) (New York: Farrar Straus & Giroux), p. xxix

10. Fitzgerald, F. S. (1920) *This Side of Paradise* (New York: Scribner), p. 26

11. Thomas, L. (1983) *The Youngest Science* (New York: Viking), p. 26ff

12. "L.T." (Thomas L.) (1931) Out Like a Light. *Princeton Tiger,* June 3

13. Synnott, M. G. (1979) *The Half-opened Door: Discrimination and Admissions at Harvard, Yale, and Princeton, 1900–1970* (Westport, CT: Greenwood), p. 168

14. Thomas, *The Youngest Science,* p. 87

15. Quiller-Couch, A. T., ed. (1919) *The Oxford Book of English Verse, 1250–1900* (Oxford and New York: Clarendon Press); Bartleby.com, 1999, www.bartleby.com/101/. Accessed Jan. 2011

16. Thomas, L. (1983) Humanities and Science. *Late Night Thoughts on Listening to Mahler's Ninth Symphony* (New York: Viking), p. 155

17. Swingle, W. W., Pfiffner, J. J. (1929) Preparation of an Active Extract of Suprarenal Cortex. *Anat. Rec.* **44**, 225

18. Swingle, W. W., Pfiffner J. J. (1930) Further Observations on Adrenalectomized Cats Treated With an Aqueous Extract of the Suprarenal Cortex. *Science* **71**, 489–490

19. Loeb, J. (1920) Chemical Character and Physiological Action of the Potassium Ion. *J. Gen. Physiol.* **3**, 237–245

20. Weissmann, G., Thomas L. (1962) Studies on Lysosomes. I. The Effects of Endotoxin, Endotoxin Tolerance, and Cortisone on the Release of Acid Hydrolases from a Granular Fraction of Rabbit Liver. *J. Exp. Med.* **116**, 433–450

21. "L.T." (Thomas L.) (1932) A Disrespectful Note on the Divine Plan. *Princeton Tiger,* Oct. 10

## Icarus and Fukushima Daiichi:
## Human Factors in a Meltdown (Sv=1J/kg.w)

1. Anonymous (1952) Body Radiation Tested; Swedish Scientist Perfects New More Sensitive Machine. *The New York Times,* Apr. 5, 5

2. Belson, K. and Tabuchi, H. (2011) Confidence Slips Away as Japan Battles Nuclear Peril. *The New York Times,* Mar. 30, 1

3. Ikaria Holidays. http://www.ikaria-holidays.com/english/index.php? categoryid=21&p2_articleid=19

4. Trabidou, G., Florou, H. (2010) Estimation of Dose Rates to Humans Exposed to Elevated Natural Radioactivity Through Different Pathways in the Island of Ikaria, Greece. *Radiat. Prot. Dosimetry* **142**, 378–384

5. Schneider, M. (2011) Nuclear Fallout Comes With Aura of Arrogance. *Business Week,* Mar. 22. http://www.businessweek.com/news/2011-03-22/nuclear-fallout-comes-with-aura-of-arrogance-mycle-schneider.html

6. IAEA Fukushima Nuclear Accident Update Log. http://www.iaea.org/newscenter/news/2011/fukushimafull.html

7. Grady, D. (2011) Radiation Is Everywhere, But How to Rate Harm?" *The New York Times,* Apr. 5, D1

8. Yamaguchi, M. (2011) Radiation Near Japan Reactors Too High for Workers. AP, Apr. 18, at 11:44 AM ET

9. World Nuclear Organization. http://www.world-nuclear.org/info/inf05.html

10. Bradshaw, K., Tabuchi, H., Pollack, A. (2011) Japan Officials on Defensive as Nuclear Alert Level Rises. *The New York Times,* Apr. 12, 1

11. Homeland Security Newswire. http://www.homelandsecuritynewswire.com/robot-reports-high-radiation-inside-crippled-reactors

12. Anonymous (2011) Japan: Radioactive Leaks into Sea Said 20,000 Times Above Limit. *BBC Monitoring Asia Pacific,* Apr. 21

13. Brenner, D. J. (2011) Fukushima's Radiation Fallout, *The Wall Street Journal,* Apr. 11, A1

14. Karolinska Institutet. Rolf Sievert, the Man and the Unit. http://ki.se/ki/jsp/polopoly.jsp?l=en&d=9498&a=18510

15. Sievert, R. M. (1932) *Eine Methode zur Messung von Röntgen, Radium, und Ultrastrahlung, Nebst Einige Untersuchungen Über Die Anwendbarkeit Derselben in Der Physik und Der Medizin* (Stockholm: Norstedt und Söner), p. 179

16. Anonymous (1932) Radium Dosage. *British Med. J.* **2**, 1062–1063

17. Allisy-Roberts, P.J. (2005) Radiation Quantities and Units—Understanding the Sievert *J. Radiol. Prot.* **25**, 97–100

18. http://www.europeandestinations.com/All_Packages/Montecatini_Terme_Vacations.aspx

19. Franson. P. A Visit to Ischia, Land of Radioactive Baths. http://www.traveltastes.com/ischia.htm

20. Bartoli, G., Carraturo, N., Gargiulo, E., Parrella, A., Santi, B. (1989) Evaluation of the Exposure Levels to Radioactivity in the Hot-Spring Environment of the Island of Ischia During a Year. *Ann. Ig.* **1**, 1781–1823

21. Hotel St. George Hofgastein. http://www.stgeorg.com/engl/thermaltherapy/

22. Tempfer, H., Hofmann, W., Schober, A., Lettner, H., Dinu, A. L. (2010) Deposition of Radon Progeny on Skin Surfaces and Resulting Radiation Doses in Radon Therapy. *Radiation Environ. Physics* **49**, 249–259

23. Therapeutic Radioenergic Hot Mineral Springs of Ikaria. http://www.island-ikaria.com/nature/springs.asp

24. Ovid, *Metamorphoses VIII* (8 CE; 1984) Daedalus and Icarus (Cambridge, MA: Loeb Classical Library, Harvard University Press), p. 405

# Index